SSADM: A Practical Approach

SSADM: A Practical Approach

Caroline Ashworth
Senior Consultant, AIMS Systems, London

Mike Goodland
Senior Lecturer, School of Information Systems,
Kingston Polytechnic

McGRAW-HILL BOOK COMPANY

London · New York · St Louis · San Francisco · Auckland · Bogotá
Caracas · Hamburg · Lisbon · Madrid · Mexico · Milan · Montreal
New Delhi · Panama · Paris · San Juan · São Paulo · Singapore
Sydney · Tokyo · Toronto

Published by

McGRAW-HILL Book Company Europe

Shoppenhangers Road
Maidenhead, Berkshire, England SL6 2QL
Telephone Maidenhead (0628) 23432
Cables McGRAWHILL MAIDENHEAD Telex 848484
Fax 0628 770224

British Library Cataloguing in Publication Data
Ashworth, Caroline, 1957–
    SSADM.
    1. Management. Information systems. Application of computer systems. Structured systems
analysis & systems design
    I. Title II. Goodland, Mike, 1953–
    658.4′038′0285421

    ISBN 0-07-707213-8

Library of Congress Cataloguing in Publication Data
Ashworth, Caroline, 1957–
    SSADM.: a practical approach.
    Includes bibliographical references.
    1. System design. 2. System analysis.
    I. Goodland, Mike, 1954–   . II. Title.
    QA76.9.S88G66 1990   004.2′1   89-13323

    ISBN 0-07-707213-8

910 CL 9432

Typeset by Mike Goodland
and printed and bound in Great Britain by Clays Ltd, St Ives plc

This book is dedicated to the memory of Laurie Robbins

Laurie was very actively involved in the selection of SSADM as the standard method for government. He was the head of the branch at the Central Computer and Telecommunications Agency responsible for the support, promulgation, and development of SSADM for a number of years.

Laurie was an inspiration to us both. He greatly encouraged us to write this book. We owe him much and he will be greatly missed.

Contents

Preface

The Structured Systems Analysis and Design Method (SSADM) is rapidly becoming the standard method for the analysis and design of information systems throughout the United Kingdom. The method was selected as the mandatory method for UK government projects in 1981 but has spread in use into the private sector and beyond into other countries. It is a very comprehensive method and, although the techniques do not require a particularly high level of skill to learn, it is often difficult to apply them correctly without guidance from an experienced user. This book is an attempt to distil some of that experience into a form which will both teach the basic principles of the method and also demonstrate how it can be applied. This is done by dividing the book into two parts.

Part 1, SSADM concepts, consists of two chapters. An introductory chapter describes the need for SSADM, shows how it fits into the systems development life cycle, and describes briefly the structure, techniques and documentation used by SSADM. Chapter 2, Three views of the system, describes the basic views that SSADM takes of an information system and explains their interrelationships.

Part 2, Working through SSADM, follows the structure of SSADM with a separate chapter on each of the six stages of SSADM. Each chapter begins with an introductory section describing the stage. The remainder of each chapter is divided into several sections each describing a step. The overall objective of Part 2 is to show how SSADM might be used on a typical project. The reader is taken through the development of the major SSADM end-products and shown how they should be developed. To illustrate the development of a SSADM project we have used a case study. This concerns the activities of a fictional vehicle rental company, Yorkies. A description of the current system is given in Appendix C. Exercises are given at the end of many of the sections. These test the techniques used in the step described in the section. Suggested answers are given in Appendix D. A glossary of SSADM terms is given in Appendix A and a bibliography in Appendix B.

Because of the approach taken here, there are many circumstances and applications of the method that are not covered. This would take up a whole series of books! Instead, it is our intention that this book should impart an understanding of the objectives of the method which can be applied to a variety of projects. It has been a great temptation to us to try to add a lot of 'ifs' and 'buts' to many of the sections as our experience has been that every project is different and requires a tailored approach. We hope that, by keeping it simple, and emphasizing the purpose of each technique and task, the basic principles can be mastered.

This book describes version 3 of SSADM and covers the syllabus set by the Systems Analysis Examination Board of the British Computer Society for the Certificate of Proficiency in SSADM. We hope that this book will be of use to professionals taking that examination and to others interested in or using SSADM. Many undergraduate and postgraduate courses are now teaching SSADM in some detail; much of the material in this book has been used on the Information Systems courses at Kingston Polytechnic

Acknowledgements and thanks are due to a large number of individuals and organisations who have helped with this book. Our respective employers during the time of writing provided support, encouragement, and facilities. This includes Kingston Polytechnic, Scicon Ltd, and AIMS Systems. The staff and students of the School of Information Systems at Kingston Polytechnic deserve particular thanks for acting as 'guinea pigs' for the case study and many of the exercises. Julian Speller, David Gooda, and Karel Riha contributed words and advice. Wendy Sharpin and Karel Riha helped with the production of the many diagrams used in the book. Several people have usefully reviewed the manuscript at various stages including Susan Keeler and Mike Hill of the CCTA, Karel Riha (again) from Kingston Polytechnic, and Darryl Ince from the Open University.

Part 1 SSADM concepts

1. Introduction

1.1 Why use SSADM?

SSADM is one of the most mature and widely used structured methods in the UK. However, it requires a significant investment in training and learning curves, so why should an organization consider taking it on? Some may not acknowledge the need for systems analysis, some may see the need, but not know why a structured method is better than traditional methods, and others may need to be convinced that SSADM has advantages over the other structured methods currently available. This section addresses all three standpoints and explains the underlying principles of the method.

Why have systems analysis at all?

It is often difficult to explain what is achieved by systems analysis and design—especially when talking to a user who wants a system tomorrow! After all, their 14-year-old son can knock together a quick program on his ZX-Spectrum in a couple of hours. Surely, a larger system is just the same but a bit bigger? Why should it take so long to design a system? And then there are the traditionalists to whom systems analysis consists of refining a design that is laid down from the start. To examine the reasons for using a method, perhaps the subject is best tackled from the point of view of something that is a comparable investment: building a house.

Imagine that you have bought a plot of land in order to build yourself a house. This is a big investment and you want to be sure that you end up with the house that you want. How do you proceed?

Start constructing it straightaway?

You could go to your local do-it-yourself superstore and buy some materials and start laying the foundations straightaway. After all, houses are pretty standard and you have recently built a shed (with some help from your 14-year-old son), so you are quite an expert. Unfortunately, you get carried away and find out when it is finished that the walls are not strong enough to support the roof and you forgot to put the electricity cables in the walls, so there is no light! The whole thing fails because a lack of planning meant that even simple standard parts of the design of a house were missed (Fig. 1.1).

Fig. 1.1

Go directly to a builder?

You look in the Yellow Pages and find a builder who seems to have the right qualifications. Surely, if you go to someone who has had a lot of experience in building houses, he won't make the same mistakes you made? You show the builder your plot of land and he agrees to go ahead and start building. You tell him you want some bedrooms, a living room, a kitchen and a bathroom, and then you go away and leave him to it. Six months later he tells you he has finished and you go to look at the house (Fig. 1.2).

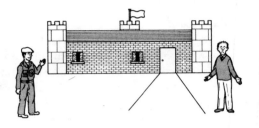

Fig. 1.2

Oh dear! The builder's tastes in house design are not very conventional and he has built you a bungalow with turrets! He has built a house that meets your instructions but there are all sorts of things about it that you hate—the rooms are small and dark whereas you have always liked large light rooms. The place is habitable but you certainly don't want to live there. You have a house that doesn't meet your requirements because you didn't tell the builder exactly what you wanted and he never came and checked with you that what he was doing was what you wanted. You assumed that because he is an expert in building a house, he didn't need to be told what to build.

Employ an architect to design it first?

You have managed to sell your monstrosity and bought another plot of land. This time you have learnt your lesson! You employ an architect. You have a series of meetings together. There are several things you tell him about your requirements. Then he goes away, draws a few plans and shows them to you to clarify a few points. To help you envisage it, he produces a few artist's impressions and non-technical drawings that you can all understand. He asks you some searching questions about how you want the rooms laid out, and makes absolutely sure he understands your requirements. Also, he points out to you that there are certain building regulations that must be adhered to so you have a good understanding of the constraints. He finally draws plans from several viewpoints and if there is anything you don't understand, he is there to explain it to you. Finally, you are happy that the plans he has drawn will meet your requirements. He asks you to authorize him to go ahead and develop the system. You sign the necessary papers and he gives the plans to a builder who proceeds to build your dream house (Fig.1.3). You are happy at last!

Fig 1.3

This rather simplistic example illustrates the need to specify requirements before construction of a house (or system) is started. Although it may seem that the requirements are fairly straightforward, constraints may be missed or interdependencies overlooked during development. A problem is far cheaper to put right early in the process than leaving it until the final day of the implementation!

How do we find out exactly what the requirements are? The future users are no more expert in expressing their needs to a programmer than the hapless landowner was in expressing what he wanted to the builder! Left to their own devices, computer people will implement the system that is most convenient to build, which will not necessarily be what the user wanted. The systems analyst takes a similar role to that of the architect—as communicator between client and builder. Some of the underlying principles of systems analysis, which are also principles of SSADM, help make sure that the user requirements are fully specified.

User involvement During the design process, the architect constantly made sure he understood the requirements by producing non-technical plans for the customer to look at and discuss. It is a basic principle of SSADM that the users have involvement in, and commitment to, the development of their system from a very early stage. By ensuring that the specification and design match the user's requirements at each stage of analysis and design, the risks of producing the 'wrong' system are very much reduced and possible problems can be sorted out before they become unmanageable.

Quality assurance The architect needed authorization from the customer to go ahead once the plans were agreed. In SSADM, formal quality assurance reviews are held at the end of each stage where the user is asked to 'sign off' the design so far. The end products for the stage are scrutinized for quality, completeness, consistency, and applicability by users, developers, and by experienced systems staff external to the project.

Separation between logical and physical specifications The requirements were expressed in logical terms first before the final architecture was known. This helped the architect to determine the best way to satisfy the requirements before going into the physical details. SSADM separates logical design from physical design. A hardware/software-independent logical design is produced which can easily be translated into an initial physical design. This helps the developers to address one problem at a time and prevents needless constraints being added at too early a stage in development. This also helps communication with users who may not be computer literate but are perfectly able to validate a logical specification or design of their system.

It is important to investigate what is required It is rare for any users to be able to describe in detail everything that is required without a lot of prompting. The architect in the example above played a vital role, using his experience of other designs and his knowledge of the techniques of planning, in asking about many of the details. Similarly, the systems analyst will need to ask the users many questions about what is required.

Why use a structured method?

Structured methods share these characteristics:

- they structure a project into small, well-defined activities and specify the sequence and interaction of these activities;
- they use diagrammatic and other modelling techniques to give a more precise (structured) definition that is understandable by both users and developers.

Why can a systems analyst not use his or her experience and just ask all the right questions? Some of the advantages of structured methods are given here.

Structured analysis provides a clear requirements statement that everyone can understand and is a firm foundation for subsequent design and implementation Part of the problem with a systems analyst just asking 'the right questions' is that it is often difficult for a technical person to describe the system concepts back to the user in terms the user can understand. Structured methods generally include the use of easily understood, non technical diagrammatic techniques. It is important that these diagrams do not contain computer jargon and technical detail that the user won't understand—and does not need to understand.

More effective use of experienced and inexperienced staff Another part of the problem with this approach is the availability of staff with enough experience to ask all the right questions. A structured method does not remove the need for experienced staff, but it does provide the option of spreading the experience more thinly. The use of structured techniques means that certain tasks can be delegated to inexperienced staff who can then be guided by the more experienced.

Improved project planning and control The use of a structured approach allows the more effective management of projects. Splitting a project down into stages and steps allows better estimation of the time taken to complete the project. Also, by following a detailed plan, it will be possible to detect slippage as it occurs and not just before the system is due to be implemented.

Better quality systems By making the specification very comprehensive it is possible to ensure that the system built will be of a high quality. The use of structured techniques has been found to lead to a system that is very flexible and amenable to change. Within SSADM, users participate in formal quality assurance reviews and informal walk-throughs and 'sign off' each stage before the developers progress to the next. This means that the analysts can be confident that the new system will meet the user's requirements before it is built.

Why choose SSADM?

As stated above, SSADM is one of the most mature methods in the UK. The experience gained in the first few years after it was introduced as a government standard has been fed back into the method to ensure its usability and practicality. Because of this, and because of the way SSADM has taken on the best ideas from other methods, it has several significant advantages over its nearest rivals.

One of the main advantages is that SSADM builds up several different views of the system which are used to cross-check one another. In the building example above, to help the customer visualize the final building, the architect drew several different representations—a cross-sectional view, artist's impressions, etc. This probably helped the architect to validate the plans as he made sure that each view was consistent with the others. In SSADM, three different views of the system are developed in analysis. These views are closely related to one another and are cross-checked extensively for consistency and completeness. The equal weight given to these three techniques and the prescriptive procedures for checking them against one another is a great strength of the SSADM approach. The three views are:

- the underlying structure of the system's data (the Logical Data Structure);
- how data flows into and out of the system and is transformed within the system (Data Flow Diagrams);
- how the system data are changed by events over time (Entity Life Histories).

Another advantage of SSADM over a number of methods is that it combines techniques into a well-established framework, and so, as well as providing the techniques for the analyst, it gives guidance on how and when to use them. Even though SSADM adopts this rather prescriptive approach, there is still a large amount of flexibility within the method and the method should be tailored to specific project circumstances.

1.2   Overview of SSADM

This section describes how SSADM fits into an overall systems development life cycle. It gives a brief overview of the structure of SSADM, and explains what each of the stages are. The major techniques of SSADM are also briefly described and the documentation standards explained.

SSADM and the system development life cycle

SSADM is used in the development of systems, but it does not cover the entire system life cycle. Figure 1.4 shows a typical system life cycle, indicating where SSADM fits into the entire procedure. We describe below what is meant by each of these phases and explain how SSADM fits into them.

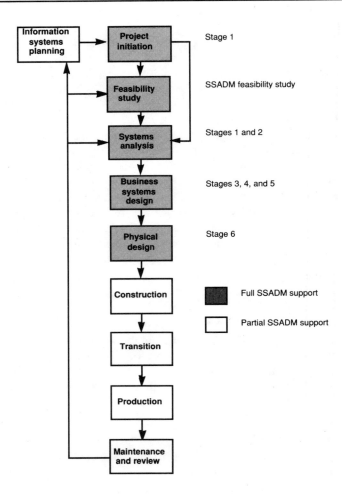

Fig. 1.4

Information systems planning

Many organizations, recognizing the contribution of information systems to their success, have invested in strategic planning for the development of future and existing information systems. Recently many methods for information systems planning have been put forward. These take a variety of approaches but generally the result from the planning exercise will be an analysis of the organization's present position, recommendations as to which systems should be developed or enhanced, a plan showing the order in which these projects should be done, and outline project plans and terms of reference for each project.

Two of the techniques used by SSADM—Data Flow Diagrams and Logical Data Structures—are used, in some form, by many of these methods. Many projects which have used SSADM have been initiated by an information systems planning study. To this extent SSADM offers partial support to this activity. There is at present no information systems planning component within SSADM, although it is likely that one will be developed for future versions of the method.

Project initiation

This is the phase where the project is set up, terms of reference agreed, team members assigned, and plans drawn up. SSADM provides detailed guidelines for this activity. In this book we develop a case study through the steps and stages of SSADM—this initiation activity is discussed in Sec. 3.2.

Feasibility study

This is the phase where it is decided whether the project is technically possible, whether it can be financially and socially justified, and whether the new system will be accepted by the organization. Feasibility studies have become less popular recently with the activity either being part of an information systems planning study or the project being a 'must have'. SSADM provides detailed guidelines on the conduct of feasibility studies detailing the steps and stages required. In this book we concentrate on 'one-pass' analysis and design, without a feasibility study. However, the techniques used are the same and the approach used by feasibility SSADM is very similar to that described in this book.

Systems analysis

Here the current system is analysed in great detail to determine the requirements for a new system. SSADM does not give guidelines on such basic systems analysis skills as interviewing and other data collection methods but provides the means of recording and analysing the results of the investigation. Stage 1 of SSADM deals with the analysis of the current system and stage 2 specifies the requirements for the new system. These two stages are introduced later in this chapter and are fully described in Chapters 3 and 4.

Business systems design

The requirements for the new system will have been broadly specified in the previous phase. In this phase various technical solutions that meet the requirements are evaluated and one selected. A detailed logical design of the new system is developed which shows clearly, in a non-technical way, how the new system will operate within the business. This phase is dealt with by SSADM stages 3, 4, and 5. They are introduced later in this chapter and fully described in Chapters 5, 6, and 7.

Physical design

The logical design is converted to a design that fits the computer hardware and software selected. This is known as the physical design and is dealt with by stage 6 of SSADM. Physical design involves the specification of files (or database definitions), the specification of programs, and the detailed operating and manual procedures that support them. This phase is introduced later in this chapter and is described in more detail in Chapter 8.

Construction

This concerns the programming, the assembly of programs into a system, and the testing of the system. SSADM does not address this phase. However, the plans for system building and testing are laid in SSADM stage 6. Many projects are now using fourth-generation environments for systems development and have integrated their use into stages 5 and 6 of SSADM; some of these ideas are discussed in Chapters 7 and 8. Prototyping has also become an important component of SSADM and this is discussed in Chapter 4.

Transition

This phase involves the transition from operating the old system to operating the new. It involves the installation of equipment, the conversion of old system data to the formats required by the new system, and the training of users. Some systems' life cycles join the construction and transition phases together to form an implementation phase SSADM does not fully address the transition phase although the plans for it are developed in stage 6.

Production

This phase begins when the system has been completely handed over to the users. The term production conveys that the system is operating and producing the information that was required of it. This activity is not supported by SSADM.

Maintenance and review

Throughout the production phase the system will require maintenance in various ways: correction of errors, adaptation to new software and hardware releases, and minor enhancements. The system will need to be reviewed to show how well it has met the requirements and objectives set for it and whether it continues to meet the users' requirements. These enhancements and reviews may lead into further system studies as shown in the diagram. Guidelines are available as to how SSADM can be employed in a maintenance environment, although these are beyond the scope of this book.

Basic principles of SSADM

SSADM is a data-driven method. This means that there is an basic assumption that systems have an underlying, generic, data structure which changes very little over time, although processing requirements may change. Within SSADM, this underlying data structure is modelled from an early stage. The representation of this data structure is checked against the processing and reporting requirements and finally built into the system's architecture.

The structured techniques of SSADM fit into a framework of steps and stages, each with defined inputs and outputs. Also, there are a number of forms and documents that are specified which add information to that held within the diagrams. Thus, SSADM consists of three important features:

- Structures   define the frameworks of steps and stages and their inputs and outputs.
- Techniques   define how the steps and tasks are performed.
- Documentation   defines how the products of the steps are presented.

Each of these features is described below.

Structure of SSADM

Figure 1.5 shows the stages of an SSADM project. These stages can be preceded by an optional feasibility phase (earlier we explained that this phase is often omitted and is not dealt with in this book). Each stage is broken down into a number of steps which define inputs, outputs, and tasks to be performed. The products of each step and the interfaces between steps are clearly defined.

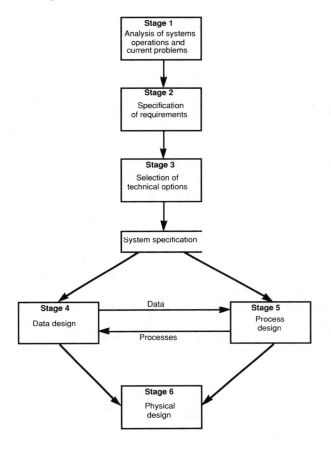

Fig. 1.5

The structure of the method illustrates several features of the SSADM approach :

1   The current system, in its current implementation, is studied first in order to gain an understanding of the environment of the new system.

2   This view of the current system is used to build the specification of the required system. However, the required system is not constrained by the way in which the current system is implemented.

3   The specification of requirements is detailed to the extent that detailed technical options can be formulated.

4   The detailed design is completed at the logical level before implementation issues are addressed.

5.   The logical design is converted into physical design by the application of simple (first cut) rules. The resulting design is tuned using the technique of physical design control before implementation.

Stage 1: Analysis of system operations and current problems
The current system is investigated for several reasons, including the following:

- the analysts learn the terminology and function of the users' environment;
- the old system may form the basis of the new system;
- the data required by the system can be investigated;
- it provides the users with a good introduction to the techniques;
- the boundaries of the investigation can be clearly set.

The third reason illustrates one of the principles of SSADM that the underlying structure of the data of a system will not change much over time. Even though the introduction of a new computer system may change the functions—a computer system can significantly increase what can be tackled by the users—the underlying data required to perform the functions will not change very much.

If there is no current system, for example where there is a new law that requires support, this stage consists of just initiating the project and beginning to document the new requirements.

Stage 2: Specification of requirements
In order that the new system will not be constrained by the current implementation, there are a number of steps within this stage to lead the analysts gradually away from the current system towards a fresh view of the requirements.

Firstly, the current system view built up in stage 1 is redrawn to extract *what* the system does without any indication of *how* this is achieved. The resulting picture is the logical view of the current system. This allows the analyst to concentrate on what functions are performed in the current system and to make decisions about what must be included in the new system.

The current system is left far behind by the Business System Options which are completed next. They reflect the different ways in which the system might be organized to meet the requirements. The decisions here are not implementation decisions (although they may constrain the way the system is implemented). Instead, this is a way of taking a fresh view of what the system is required to do and how the system can be organized to meet the underlying business objectives.

Based upon the selected Business System Option, a detailed specification of the required system is built up and checked extensively.

Stage 3: Selection of technical options
At this stage, if the purchase of new computer equipment is required, the development team have enough information to compile the different implementation options for the system. Each option is costed out and the benefits weighed against the costs to give the user some help in choosing the final solution. This might form the basis for selecting the final system hardware.

Stage 4: Logical data design

This stage builds up the logical data design so that all the required data will be included. It applies a relational analysis technique to groups of data items in the system to act as a cross-check on the data definition built up in stage 2.

The final data design is checked against the logical processes, developed in stage 5, to be sure that all the data needed by the processes are present in the data design.

Stage 5: Logical process design

The definition developed in stage 2 is expanded to a very high level of detail so that the constructor can be given all the detail necessary to build the system. This processing definition is checked against the data definitions derived in stage 4.

Stage 6: Physical design

Here, the complete logical design—both data and processing—is converted into a design that will run on the target environment. The initial physical design is tuned on paper before being implemented so that it will meet the performance requirements of the system.

In this stage, much of the documentation required during the construction and transition phases is produced.

Other templates for SSADM

We have described above and concentrate, in this book, on the basic structure of SSADM. However, every project is different and to a certain extent requires its own tailored method. Towards this end, different SSADM structures, or templates, have been developed for common project circumstances. These include templates for use with application packages, for microcomputer systems, for use in a maintenance environment, and 'fast path' templates for use when a solution is required 'yesterday' and risks have to be taken. These templates use the SSADM techniques and principles described in this book; their differences with basic SSADM will be structural in that certain steps are omitted or added and that the order of steps will be different.

Structured techniques

The techniques of SSADM give standards for how each step and task is to be performed. The rules of the syntax and notation of each technique are supplemented with guidelines on how they should be applied in a particular step. The diagrammatic techniques of SSADM are:

- Data Flow Diagrams;
- Logical Data Structures;
- Entity Life Histories;
- Logical Dialogue Design.

In addition, there are techniques and procedures that are not diagrammatic including:

- relational data analysis;
- first cut rules;
- physical design control;
- quality assurance;
- project estimating.

This book gives clear guidelines on each of the techniques and shows how they are inter-related and can be used to cross-check one another. The principal diagrammatic techniques and procedures are described below and explained in more detail in subsequent chapters.

Logical Data Structure (LDS)

This is a method for describing what information should be held by the system. The approach used in SSADM is very similar to entity modelling in other methods. A diagram is produced showing the entities and their relationships, this is further documented by a set of entity description forms detailing their data contents.

A Logical Data Structure is produced for the current system. This is extended to meet the requirements of the new system, resulting in a Required System Logical Data Structure. This is converted into the Composite Logical Data Design by comparison with the results of relational data analysis. The Composite Logical Data Design is used as the basis for the physical data design.

Data Flow Diagrams (DFDs)

Data Flow Diagrams are a widely used technique for representing the information flows of a system. The approach used in SSADM is similar to that described by DeMarco (1979), although a different diagrammatic notation is used. The diagrams represent the external agents sending and receiving information; the processes that change information; the information flows themselves; and where information is stored. (They should really be called 'Information Flow Diagrams'.) The diagrams are hierarchical in nature with a single, top-level diagram decomposing to many lower-level diagrams, each representing different parts of the system.

Data Flow Diagrams are used in the early stages in systems analysis to help understand the present system. As the project proceeds they are used to represent the required system and are further used as the basis for program specification.

Entity Life Histories (ELHs)

These are models of how the system's data is changed over time by events acting on entities. For each entity the sequence, selection and iteration of events affecting it are shown using a notation derived from Jackson (1975).

An event is whatever triggers a process to update system data. As it would be too complicated to model the entire set of events for a whole system at once, the effects of the events upon each entity from the Logical Data Structure are modelled. These individual views of the event sequences are drawn together in an entity/event matrix (ELH Matrix) and Process Outlines.

Logical Dialogue Outlines

Logical Dialogue Outlines were introduced into SSADM to allow developers to specify requirements for man–machine dialogues at an early stage in the development. The prototyping of dialogues using a screen painter, or similar rapid development systems software, to demonstrate the man–machine interface to users, is obviously more effective in the specification of user requirements for dialogues, so dialogue outlines are designed to be used generally where prototyping facilities are not available. Logical Dialogue Outlines are produced for any complicated on-line processing, the technique being used towards the end of requirements definition in stage 2. The data items flowing across the man–machine boundary are detailed, the sequence of logical 'screens' defined, and an overview of the processing necessary to perform the dialogue modelled by using a flow-chart style notation. It is also possible to add the requirements for the time taken at each stage of the dialogue, points at which users will be required to make decisions, an indication of some messages that might be used, and a cross-reference to operations on Process Outlines.

Relational data analysis

Relational data analysis, based upon relational theory, is used in the logical design stage of SSADM (stage 4) where it complements the logical data structuring done during requirements analysis. The merging of the two techniques results in the Composite Logical Data Design which is the basis for physical database or file design.

Any collection of data items that have been defined without direct reference to the Logical Data Structure can be used as an input to relational data analysis. Most commonly, the Input/Output Descriptions or screen definitions are used as inputs to this technique.

Relational data analysis consists of a progression from the original, unnormalized, data through several refinements (normal forms) until the data items are arranged to eliminate any repeating items or duplication. The results of performing this analysis on several different groups of data items are merged, or optimized, to give sets of data items that should correspond to the entities on the Logical Data Structure. At this point, the Logical Data Structure is merged with these results of the data analysis.

The process of relational data analysis ensures that all data items required by the system are included in the system's data structure. Also, it is a good way to ensure that the data is fully understood. Although the rules of relational data analysis appear to be mechanistic, to apply them effectively the underlying relationships between data items must be well understood.

First cut rules and physical design control

The conversion of the logical process and data design into a workable physical design takes place in two phases. First, simple rules are applied which crudely convert the logical design into a corresponding design for the target environment. This design might work, but would probably not be very efficient or exploit the features of the particular hardware or software that will be used. Therefore, the 'first cut' design is tuned using a process called physical design control. This consists of successively calculating the performance of the system (time taken to execute certain critical transactions, space requirements, and recovery times), then modifying the design slightly and recalculating until the performance objectives (defined in stage 3) are met.

Quality assurance

SSADM lays a great emphasis on holding formal quality assurance reviews at the end of each stage. It is very important to ensure that the products from each stage are technically correct and that they meet the objectives of the users. The work for the second stage of SSADM has its foundations in the work done in the first stage. This principle applies throughout the project; each stage builds on the work done in the previous stage.

Obviously, with poor foundations, there is a high risk that all subsequent work will be poor. The approach to quality assurance is not discussed elsewhere in this book so this subject is discussed in more detail than the other SSADM techniques in this section.

A formal sign-off by a group consisting principally of users emphasizes the joint responsibility for the project of both the users and the project team. This ensures the continuing active interest of the users in the project and avoids the situation commonly encountered in systems development when minimal communication between the project team and the users leads to an implemented system that does not meet the users' requirements.

Products from each stage should be reviewed by a team comprising responsible users who will have the authority to authorize the continuation of the project and at least one person with a good understanding of SSADM who will be referred to here as the 'technical reviewer'. This should be done on a formal basis, to force the correction of errors identified by the reviewers before work is allowed to proceed to the subsequent stages.

The following procedures are an example of how a quality assurance review might be undertaken within SSADM.

Before the review All participants receive an invitation to the review meeting one week in advance of the meeting, together with a neat copy of all the documents they will be required to review. If any of the reviewers is unfamiliar with the conventions of the diagrams, then the analysts might arrange to explain the aspects of the diagrams that are relevant to a reviewer. This can be done on a one-to-one basis, but can be achieved more efficiently, when a number of people are involved, by organizing a presentation to state the purpose and basic conventions of the diagrams with a more general discussion about quality assurance review procedures.

The review meeting The actual review would not be more than one to two hours long. The chairman is either a user who has been closely involved with the project, or the project team manager. The meeting begins with the circulation of an agenda, possibly some introductions, and a clear statement of the objectives of the meeting. The meeting should not attempt to solve the difficulties that might arise, but simply to highlight errors for subsequent resolution away from the meeting. An analyst from the project team walks through the documentation being reviewed and invites comments from the reviewers. A list of errors is compiled by the chairman and agreed by the meeting. The reviewers may decide that the documentation contains no errors and meets its objectives, in which case they will sign the stage off at this meeting. More commonly, there will be a number of non-critical errors detected, in which case the documentation may be signed off provided that certain follow-up action is taken and subsequently agreed by the reviewers out of the meeting. If the number of errors is very great, and the reviewers are not confident that the project team have met the objectives of the stage, then a date for another quality assurance review is set and the documentation failed.

After the review Any necessary corrections are made to the documentation within a week of the review and circulated to the members of the review team. If the errors are only minor, the reviewers may sign it off individually. If the errors are more severe, the documentation is reviewed a second time at another review meeting.

The resources required to hold a quality assurance review are significant and should not be underestimated when the project plan is being prepared. At least three elapsed weeks should be allowed for each formal review, and one to two weeks for informal reviews. It is a temptation to cut this time when project time-scales are tight. But compared to the weeks or months that might be wasted later in the project on trying to sort out compounded errors arising from poor quality assurance, it is time well spent.

Project estimating

Project estimating guidelines are based upon the techniques, steps, and stages of SSADM. Certain factors will make time-scales longer or shorter, for example the number of user areas and the complexity of the project. The estimating guidelines are applied after an initial Data Flow Diagram and Logical Data Structure have been drawn. The number of processes and entities on these initial diagrams give an indication of the number of diagrams that will be completed throughout the project.

The results of the estimating guidelines are refined throughout the project. The estimates produced at the beginning of a project will not be accurate but will give some idea of the order of magnitude of a project.

Documentation

Documentation standards define how the products of this development activity should be presented. Several different kinds of document will be produced during a project: diagrams, forms, matrices, and narrative reports. Working documents are produced as a means of developing some of the diagrams. Other documents are part of the formal documentation standards of SSADM and are carried forward into later steps of the project. Some of these will need to be maintained throughout the system's life cycle.

An important part of the diagrammatic techniques are the standards that govern their layout, symbols, and content. This standardization facilitates understanding of the diagrams within the project team, with users, and with others interested in the project.

In addition to the diagrams there is a considerable amount of highly structured information that needs to be developed. Some of this is necessary to support the diagrams, for instance each data entity shown on a Logical Data Structure diagram needs to be defined, and its constituent data items must be specified. Some of this structured information is not directly related to the diagrams, for example a Problems/Requirements List is developed which describes and gives a severity weighting to each problem or requirement identified in analysis. To help record this structured information a number of standard forms are suggested. We have used these forms throughout the book, although it should be emphasized that these forms can be modified for specific projects.

In addition to forms, matrices are used to help start some of the diagrams. An entity matrix is used to identify relationships between entities as an initial step in the logical data structuring technique and an entity/event matrix is used as a basis for Entity Life Histories. Matrices are also sometimes useful in cross-checking one technique with another.

Often formal reports such as feasibility study reports or full study reports will be required. SSADM provides no detailed recommendations on the content or format of such reports; these will often be set by the organization. However, much of the SSADM documentation, both diagrams and forms, will need to be integrated into these formal reports.

Finally, although the emphasis of this book is in describing a paper-based method, we should stress that very few projects will use only manual methods of documentation. There are now many computer-aided systems engineering (CASE) tools on the market. Some of these support SSADM and some support some of the SSADM techniques. These tools are invaluable in producing high-quality documentation, in checking consistency and completeness, in enabling rapid amendment of diagrams and other structured information, and in many other ways. We have chosen to present the method as a manual one for two main reasons. First, because we wished to concentrate on the method itself rather than its software implementations. Secondly, the market is changing very rapidly and our use of any particular product or range of products could soon be out of date.

2. Three views of the system

SSADM takes three basic views of an information system.

- Logical Data Structures — show what information is stored and how it is interrelated;
- Data Flow Diagrams — show how information is passed around;
- Entity Life Histories — show how information is changed during its lifetime.

Each of these views is developed through systems analysis (stages 1–3) and logical design (stages 4 and 5) before conversion to an executable physical design. In each part of SSADM all three of these views are used and interrelated. It is therefore important to gain a good understanding of each view and how they fit together as a whole. In this chapter we discuss each view separately; concentrating on the notation and the conceptual information represented. In the final section of the chapter we describe how the views interrelate and can be cross-checked against each other, some readers may find it easier to skim this section and return to it after seeing how the different views are used in practice in Chapters 3 and 4.

Note that we do not attempt to describe how the different views are developed in this chapter. This is explained in detail in the rest of the book where we take a case study through all the stages of SSADM.

2.1 Logical Data Structures

Introduction

Computer systems are much better than human beings at storing and retrieving large amounts of data. This characteristic has led to the development of large, computer-based information systems. All organizations of any size possess vast quantities of information which may be stored in a number of ways: in people's heads, in filing cabinets, on card indexes, on microfiche, and on computer-readable media. To design computer-based information systems for these organizations it is necessary to know exactly what information should be held by the computer and to specify how that information should be organized.

SSADM makes a fundamental split between a logical and a physical view of data. The physical view represents how the system's data is held in particular technological environments. For example, if the system was currently implemented using many card indexes and paper-based files, then the way its data would be organized to maximum efficiency would be very different from the same system implemented using serial and index sequential files or using a relational database. The logical view of the system's data is completely independent of how it is implemented; it could be physically using any file or database organization or manually based file organization.

The result of the first five stages of SSADM is a complete specification of the logical data design and the logical processing design. To achieve this logical data design SSADM uses two techniques: logical data structuring and relational data analysis. Thus two separate views of the system's data are derived. These views are combined to create the complete logical data design (known as the Composite Logical Data Design).

Relational data analysis is a 'bottom up' technique where the smallest meaningful components of information—the data items—are rigorously analysed to produce a flexible complete organization of data. Logical data structuring is a 'top down' technique where things about which we might want to hold information are modelled.

Logical data structuring is a very widely used mature technique. Originally devised for database design, it is now also used in a number of systems development methods. The terms entity modelling, data modelling, and entity relationship modelling are all used by different methods to describe a similar approach to SSADM's logical data structuring.

In this section we describe first the components of a Logical Data Structure: entities, relationships, and data items. The Logical Data Structure diagram is then described in terms of its layout, validation, and supporting documentation.

Entities and data items

A Logical Data Structure of a system models the entities and their interrelationships.

Definition of an entity

An entity is something of significance to the system about which information is to be held.
Some examples of information systems and their entities are:

- Hospital information   patients, wards, diseases.
- Library systems   books, borrowers, loans, reservations, fines.
- Personnel systems   employees, jobs, grades, skills.
- Banking systems   customers, bank accounts, loans.

Type and occurrence

In each of the above examples we have described the entity type that represents a number of entity occurrences. For example, in the bank the entity type Customer represents all the occurrences of the Customer entity—Maurice Moneybags, Percy Penniless, etc.

It is important to make this distinction between type and occurrence as it often causes confusion.

Entity	*Entity*
occurrence	
Customer	M. Moneybags
	P. Penniless

In this book we always use entity to refer to the entity type and always use entity occurrence to refer to a specific entity occurrence. When we are referring to the entity type the name is begun with a capital, as in Customer.

Representation of entities

An entity is represented as a box on a Logical Data Structure diagram with the name of the entity inside it as shown in Fig. 2.1. The singular form is used to emphasize that it is the entity type rather than the number of occurrences that is being represented.

Fig. 2.1

Real world entities and system entities

The definition of an entity began: 'An entity is something of significance to the system...' A real world 'something'—a book, a person, a job—may be important to a number of different systems. The real world Maurice Moneybags may be a system entity occurrence in a number of systems: patient in the hospital example, borrower occurrence in the library, employee occurrence in the personnel, and customer occurrence in the bank. In logical data structuring we ignore the real world aspects of entities and concentrate on things about them that are important to the system. It is therefore system entities rather than real world entities that are described in Logical Data Structures, although the distinction is rarely made. The real world existence of these entities is a convenient handle which enables us to get to grips with their importance to the system being considered. Thus customers, bank accounts, loans are all real world things of significance to the banking system.

Real world attributes and data items

'An entity .. about which information is to be held': when designing information systems we need to say precisely what that information is. The real world Maurice Moneybags has a name, a wife called Joyce, a house (at Acacia Avenue) a Rolls Royce car, blue eyes, three cats and a credit card. These could all be thought of as real world attribute occurrences of the attribute types, name, wife's name, home-owner or not, home address, make of car, colour of eyes, number of cats, credit card holder or not. However, for the banking system we are not interested in the colour of our customers' eyes or the number of cats they own but in their names and home addresses. So for each system entity we can define a set of system attribute types. In SSADM these system attribute types are known as data items.

Definition of a data item

A data item is the smallest discrete component of the system information that is meaningful.

Other approaches to data modelling sometimes use the terms 'attribute', 'logical data' 'item', and 'data element' to mean much the same thing.

When a data item is implemented in a file-based computer system it is usually referred to as a field. Different database management systems use a variety of terms to refer to their implementation of data items.

Data items are not only important in logical data structuring in SSADM, the contents of data flows and the contents of the system inputs and outputs are also described in terms of their constituent data items.

In the early steps and stages of SSADM only the major data items for each entity may be loosely described; however, before physical design can commence each data item should be fully described by a description, a format, sizes (average and maximum), and by validation information.

Identifiers

It must be possible to identify uniquely each entity occurrence. For each entity this means that there must be one data item or group of data items whose value determines the value of all the other items in the entity. In other words if the value of this key data item is known it is possible to find the values of all the other data items in that entity. For example, let us take an occurrence of the entity Bank Account in Cash & Grabbs Bank:

Account No.	1234
Date Opened	12 June 1978
Branch	The Strand, London
Current Balance	£20 000

The Current Balance clearly cannot uniquely determine the values of the other data items as there obviously may be several accounts with balances of £20 000. In a similar way neither Date Opened nor Branch could determine the other values. The only data item that could uniquely determine the values of the other data items is Account No. This is then referred to as the primary key. Through the primary key we can find all of the data associated with a particular entity occurrence.

Sometimes a combination of data items is required as the key, for example if each branch of the Bank set their own Account Nos., so that it would be possible for the same account number to occur at another branch. It would then require a combination of Account No. and Branch to identify uniquely each Bank Account occurrence.

This definition of keys is dealt with in much more detail in the section on relational data analysis. The important thing to remember here is that each entity occurrence must be capable of being uniquely identified, and it is through the identifying data items that we have immediate access to all the data for the particular entity occurrence.

Relationships

Real world entities have relationships with other real world entities. Maurice Moneybags has *borrowed* the books *How to be a Millionaire* and *Tax Saving Guide* from the library, *is treated* in Ward 10 of the hospital, *has* a grade of Senior Accountant at work, and *has* a current bank account and a deposit bank account. These are all examples of real world relationship occurrences that are also system relationship occurrences in the library, hospital, personnel, and banking systems. Relationships are normally described as verbs (borrowed, treated, has) and entities as nouns (borrower, book, patient, ward, etc.). As with entities and data items, relationship type is always referred to in this book as relationship and relationship occurrence is always referred to as relationship occurrence.

Definition of a relationship

A relationship is an association between two entities that is important to the system.

Relationships are important when designing information systems because they define access from one entity occurrence to another. Thus a relationship between say patient and ward implies that from an occurrence of patient, say Mr Moneybags, we can find the occurrence of the ward in which they are treated, Ward 10. Similarly we can find which patients are in a particular ward by going from the ward occurrence to the patient occurrences.

Representation of relationships

The relationship described above is a one-to-many relationship in that *one* occurrence of ward may have *many* occurrences of patient associated with it. We could show diagrammatically the relationships between a ward and its patients (see Fig. 2.2).

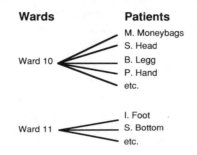

Fig. 2.2

These occurrence diagrams are useful for explaining ideas but a more concise notation is required for specifying systems (see Fig. 2.3).

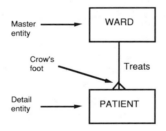

Fig. 2.3

Figure 2.3 shows entity types and relationship types. The line with the crow's foot describes the relationship. The crow's foot is always shown at the 'many' end. The entity at the one end is often referred to as the master entity and the entity at the 'many' end referred to as the detail entity.

Naming relationships

In SSADM relationships are only normally named when more than one relationship exists between the same pair of entities (these are referred to as multiple relationships and are discussed later).

Complex Logical Data Structures may contain more than 50 entities and 100 relationships. It often happens that after much discussion a relationship is drawn, and then two weeks later the discussion will be reopened because no one can remember why the relationship was added. For this reason naming of relationships on Logical Data Structures is a good idea. If relationships are not named on the diagram they can be documented separately as Relationship Descriptions (described later in this chapter). To name a relationship it is best to use the verb of a sentence that describes the relationship. Thus 'a ward *treats* patients', 'a borrower *borrows* books', and 'a customer *has* bank accounts'. Obviously sentences can be constructed in both directions, with the subject entity in one sentence being the object entity in the other sentence, e.g. 'patients *are treated in* wards'. Some methods recommend naming the relationship at both ends, but since these names are usually only reflections of each other, and because they clutter the diagrams, we recommend that only one name is used. The direction in which the name is given should follow from the simpler construction of the sentence. Thus *treats* is preferred to *are treated in*.

The degree of relationships

The relationship between Ward and Patient was a one-to-many relationship in that at any one time a patient could only be treated in one ward, and that in any one ward there could be many patients treated. These one-to-many relationships between two different entities are the most common type occurring on Logical Data Structures. Between two entities A and B there are four possible degrees of relationship:

1. One A can be related to many B's.
2. Many A's can be related to one B.
3. Many A's can be related to many B's.
4. One A can be related to one B.

1 and 2 are examples of one-to-many relationships discussed before. Many-to-many relationships and one-to-one relationships are discussed below.

Many-to-many relationships

Example 3 above is a many-to-many relationship. Logical data structuring does not recognize many-to-many relationships and they are never shown on the diagrams. An example will make clear why this is so.

Consider the relationship between a patient and drugs in the hospital.

A patient may be prescribed many drugs and a drug may be prescribed to many patients (Fig. 2.4).

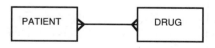

Fig. 2.4

So we have a many-to-many relationship. However, when we investigate this relationship more deeply it becomes apparent that there is information associated with the relationship 'prescribed'. In fact the more we think about it the more obvious it becomes that 'Prescription' is itself an entity, containing information such as the date of prescription, the dosage, and strength. Further analysis shows that a prescription can only be for one patient, although each patient may have many prescriptions. Hospital rules state that a prescription can be for one and only one drug and obviously a drug may appear on many prescriptions (Fig. 2.5).

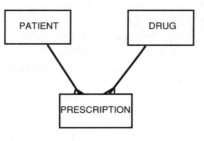

Fig. 2.5

Thus the many-to-many relationship can be broken into two one-to-many relationships. It is normally the case that what appear to be many-to-many relationships can, on further analysis, be broken into new entities and relationships. This is the main reason why many-to-many relationships are not allowed on Logical Data Structures. (It is also true that very few database management systems can implement many-to-many relationships—this is not really a reason for not allowing them: physical implementation should not dictate to logical models.)

There are rare cases where true many-to-many relationships exist. When this occurs a link entity and two one-to-many relationships are created. An example of this comes from the banking system where a customer can have many bank accounts and a particular bank account may be owned by several customers. The link entity could be called 'Participation' or 'Signatory' but this is a true many-to-many relationship as there is no information associated with it. If no suitable name could be found then the link entity could have been called 'Customer/Bank Account link' (see Fig. 2.6).

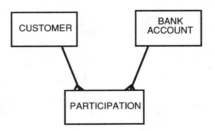

Fig. 2.6

It is important to note that the 'many' does not necessarily mean 'lots of'. Most people will only have one bank account and most bank accounts will only be owned by one person. So there will, in most cases, be only one occurrence of participation for each customer. What is being represented in the Logical Data Structure is the general case.

One-to-one relationships

This is where one A is related of only one B. Like the many-to-many relationships true one-to-one relationships rarely exist and are not shown on Logical Data Structures. An example will make this clear.

Consider the entities Borrower and Book in the library example; further thought might lead us to another entity, Loan. Examining the relationships between these three entities it becomes clear that the relationship between Book and Borrower is maintained though the Loan entity (Fig. 2.7).

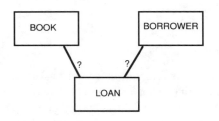

Fig. 2.7

This library is a very poor one and has only one copy of each book—this makes the example more straightforward. A borrower occurrence may at one time have several loan occurrences. Each of these loan occurrences can only be associated with only one borrower occurrence. So there is a one-to-many relationship between Borrower and Loan.

A loan occurrence can only be for only one book occurrence and a book occurrence can only have, at any one time, one loan occurrence associated with it (in this case we only need to record the current loan). So we have a one-to-one relationship between Book and Loan (Fig. 2.8).

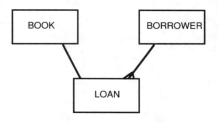

Fig. 2.8

We can, however, amalgamate the entities Book and Loan, adding to the data items associated with Book (e.g. title, ISBN, author, date of publication) the data items associated with Loan (e.g. date borrowed, date due back). This could be represented as shown in Fig. 2.9.

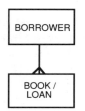

Fig. 2.9

The existence of relationships

We have discussed the degree or order of a relationship and now turn to describing how whether a relationship always exists or not is represented.

Consider a one-to-many relationship between entities A and B. There are four possible states in which a relationship could exist between these two entities.

1. *Mandatory at both ends of the relationship* If an occurrence of *A* exists it *must* always be associated with at least one occurrence of *B*; and conversely, if an occurrence of *B* exists it *must* always be associated with one occurrence of *A*.

2. *Optional at the master end and mandatory at the detail end* If an occurrence of *A* exists it *may* be associated with one or more occurrences of *B*; and if an occurrence of *B* exists then it *must* be associated with one occurrence of *A*.

3. *Mandatory at the master end and optional at the detail end* If an occurrence of *A* exists it *must* always be associated with at least one occurrence of *B*; and if an occurrence of *B* exists it *may* be associated with one occurrence of *A*.

4. *Optional at both ends of the relationship* If an occurrence of *A* exists it *may* be associated with one or more occurrences of *B*; and if an occurrence of *B* exists it *may* be associated with one occurrence of *A*.

States 1 and 2 above are not distinguished in SSADM; relationships which are mandatory at the detail end are shown by the normal crow's leg (see Fig. 2.10). In other words, in a normal one-to-many relationship the 'many' includes zero.

States 3 and 4 above are not distinguished in SSADM; relationships which are optional at the detail end are shown by the circle on the crow's leg (see Fig. 2.10). These states are referred to as *optional relationships* in SSADM. In other words, an optional relationship indicates that an occurrence of the detail can exist without its corresponding master.

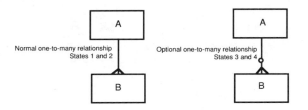

Fig. 2.10

Exclusive relationships

This is when the existence of one relationship precludes the existence of another. In Fig. 2.11 an occurrence of *B* can be owned by *either* an occurrence of *A or* by an occurrence of *C* but *never* by *both*. This is represented by drawing an arc linking the two relationships that exclude each other.

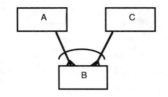

Fig. 2.11

Individual arcs, linked by identifiers ('a's in Fig. 2.12), are sometimes used when the drawing of a continuous arc would cause crossing lines or otherwise confuse the diagram.

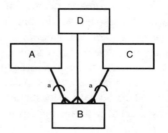

Fig. 2.12

Figure 2.13 shows that a patient occurrence can be either treated in a ward or in an outpatient department but not in both.

Sometimes exclusivity can occur across several relationships (e.g. the patient could be treated in a ward, an outpatient department, or by a community nurse)—these are shown using the same notation.

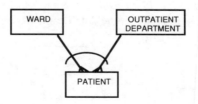

Fig. 2.13

Recursive relationships

Previously we have discussed the relationship of entity occurrences of one type with entity occurrences of a different type. Often entity occurrences have direct relationships with other entity occurrences of the same type.

Consider the personal system in which Maurice Moneybags is an entity occurrence of the type Employee. Maurice has a manager, J. Pounds, who is also an occurrence of Employee, Maurice also manages I. Shilling and A. Penny. An employee can manage many employees but is managed by one and only one employee. Thus the Employee entity has a one-to-many relationship with itself (Fig. 2.14). This recursive relationship is shown by the 'crow's leg' looping round—often referred to as a 'pig's ear'. This kind of relationship is very common in administrative systems.

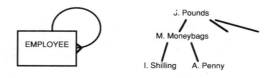

Fig. 2.14

Entities also may appear to have many-to-many relationships with themselves. When this happens a double-lobed pig's ear (sow's purse) is not drawn. Instead, as in other many-to-many relationships, a link entity is created (Fig. 2.15).

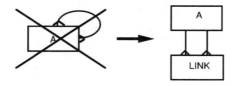

Fig. 2.15

Consider a delivery system to supermarkets; if one product is out of stock then alternative products can act as substitutes. So each product can be substituted by many other products and in turn act as substitute for many other products (see Fig. 2.16).

Fig. 2.16

Thus it appears that the Product entity has a many-to-many relationship with itself. A link entity is therefore created (Fig. 2.17). On closer inspection it is clear that there are data items associated with the link entity such as the price adjustment required and the equivalent quantity.

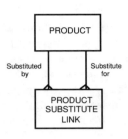

Fig. 2.17

Multiple relationships

Relationships are not always named in Logical Data Structures but are named when there are multiple relationships between the same pair of entities. It is not always obvious what the relationships are and it is helpful to distinguish them.

Multiple relationships are quite common and are often a convenient way of avoiding creation of unnecessary link entities.

Suppose we are trying to model the relationship between teams and fixtures in a sport such as football. Each team may have many fixtures and each fixture involves two teams. So it appears that we have a many-to-many relationship for which we can create an artificial link entity: Participation (see Fig. 2.18).

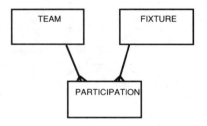

Fig. 2.18

However, a better way of modelling this is to say that as only two teams can be involved in a fixture then Fixture has two relationships with Team (Fig. 2.19).

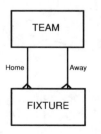

Fig. 2.19

A very frequent occasion when a multiple relationship occurs is when the movement of something is being modelled (see Fig. 2.20). An example like this occurs in the case study.

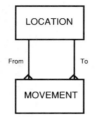

Fig. 2.20

The Logical Data Structure Diagram

Previous sections have dealt with the components of a Logical Data Structure and their relationships. This section deals with the diagram as a whole.

Layout (or presentation)

For a very large and complex system these diagrams can contain more than 100 entities and more than 200 relationships. Even simple systems have have 20 or 30 entities. It is important that these diagrams are presented in a way that makes them easy to read and understand.

Crossing Lines If at all possible the diagram should be drawn so that relationship lines do not cross. This can often be easily managed by rearranging the position of entities, as shown, for example, in Fig. 2.21.

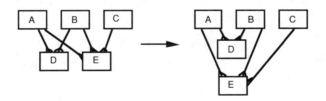

Fig. 2.21

Sometimes it is necessary to place a kink or 'dog leg' a line to avoid crossing an entity or another line (Fig. 2.22).

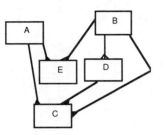

Fig. 2.22

Some practitioners and some computer-based tools show only orthogonal relationship lines (Fig. 2.23). This makes the diagrams look neat but in the opinion of the authors makes it harder, in complex diagrams, to sort out which entities are related to which others.

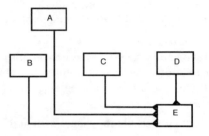

Fig. 2.23

If it is impossible to avoid crossing lines then the neatest way of drawing th s to build a 'bridge' over which one of the relationship lines passes. This is shown in Fig. 2.24.

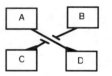

Fig. 2.24

Positioning the entities Some complex Logical Data Structures may cover several areas of the organization. Entities that fall into the same business area should then be drawn physically close to each other. This will often happen anyway because their relationships will tend to be with other entities in the same business area.

There are several advantages of presenting the information in this way:

- it is easier to understand;
- it is easier to present, the diagram can be presented in a top down fashion with particular business areas being discussed separately;
- it is easier to partition the project, with different people or teams having responsibility for different business areas.

A convention of Logical Data Structures is that they are drawn so that the 'crow's feet' are placed so that the crow stands on the detail entity (see Fig. 2.25). This makes it easy to view the diagram since all master entities are above their detail entities. This convention cannot always be followed—sometimes it will lead to crossing lines or placing entities out of context.

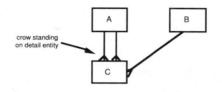

Fig. 2.25

Validation

The Logical Data Structure must be able to support the processing requirements of the system. Every bit of processing defined on the Data Flow Diagrams and every retrieval documented must be capable of being performed on the Logical Data Structure. This means that the data items required for the processing must be included in entities and that the relationships exist between those entities to enable the navigation from one entity occurrence to another for that bit of processing to be performed. This validation will be discussed further in the next section which deals with the relationship between the Data Flow Diagrams and the Logical Data Structure and in the worked case study.

Operational masters

Previously we have described two ways of obtaining access to the data in an entity occurrence:

- direct access using the primary key of the entity;
- access to the entity occurrence from another entity occurrence through a relationship.

Often validation of the Logical Data Structure tells us that, for a particular requirement, access is required on a data item other than the primary key data item. Suppose that details are required on all accounts held at a particular branch of Cash & Grabbs Bank. The entity Bank Account contains the following items:

Account No., Date Opened, Branch, Current Balance

Cash & Grabbs have 100 000 accounts, so to find the 5000 accounts based at a particular branch it would be necessary to read through all the Bank Account occurrences. So another method to access data is simply to read through serially until the data item value in the entity occurrence matches with the required one.

With this example it would be far quicker to find the accounts if there was a way of directly accessing those with the branch required. In most database management systems there are ways of doing this using secondary keys, secondary indices, or by creating new record types. In SSADM we create an operational master showing the data item that we require direct access on. This is shown in Fig. 2.26 for the banking example.

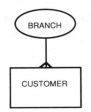

Fig. 2.26

The operational master is usually shown as an oval (as in Fig. 2.26) although sometimes a rectangle with a circle in the top left corner is used. This always means that entity occurrences related to the operational master can be accessed directly by the data item named in the oval.

Supporting documentation—Entity and Relationship Descriptions

The Logical Data Structure diagram is a summary of the data that the system must hold. It gives the groupings of data items and their interrelationships. To document fully the system's data it is necessary to go into further detail on each entity, each relationship, and each data item that has been included in an entity. This can either be presented using the standard SSADM forms or using some computer-based support tool—in either case the information held is the same.

Entity Descriptions

Consider the entity Customer in the banking example; to describe fully everything about this entity we would need to hold the following information:

Entity identifier It is useful, particularly with complex data structures, to identify uniquely each entity by using a numeric code. This helps when modifications are made to the Logical Data Structure and it is very convenient to use a short name when computer-based tools are used.

Definition or description It is important to clarify exactly what is meant by the name of the entity. Often people in other parts of the organization may have a different understanding of the same word (homonyms), e.g. the marketing department might think of a customer as someone who has expressed an interest in opening an account, whereas the rest of the bank would only regard someone as a customer when they had opened an account. In a similar way synonyms can also occur—another section of the bank might refer to a customer as a client.

By formally describing or defining every entity there is a better chance that these kind of ambiguities can be identified and resolved.

Master and detail entities These are the entities that act as masters (at the top of the 'crow's legs' attached to Customer) to the Customer entity and the entities that act as details (those at the bottom of the 'crow's legs' attached to Customer). This information is also on the Logical Data Structure but it is also useful to have a complete, separate view of each entity.

Volumetric information In order to calculate how much storage is required for the system it is necessary to know the average and the maximum number of occurrences for each entity.

Data items A list of the data items that make up the entity must be included. Every data item may be described fully in a separate Data Catalogue. On the standard SSADM Entity Description form described for each item is its *format* (e.g. numeric, alpha, date, etc.), *size* (the number of characters or bytes occupied by the data item), and any *comments* (about why or under what conditions the data item is included in this entity).

Notional keys A data item or group of data items that uniquely define an occurrence of an entity is known as its primary key. Relational data analysis is used in stage 4 of SSADM to define formally the keys of every entity. However, it is useful to nominate keys earlier, when developing the Logical Data Structure, because it makes it easier to validate the Logical Data Structure against the processing requirements. To indicate that a data item is a notional primary key (or part of one) it is marked with a tick.

Data item descriptions

Data items are the fundamental unit of data in a system; entities are made up from them, Data Flow Diagram processes transform them, inputs to and outputs from the system contain them. Each data item is separately described in a Data Catalogue. This could be held manually as a card index, as a microcomputer file, on a data dictionary package, or on a computer support tool. There is some considerable advantage in automating the maintenance of the Data Catalogue in that there will be many different data items in a system. Typically a medium-sized project (4–5 man-years for analysis and design) may have have over 400 different items and a large project may have 1000 plus. The Data Catalogue is invaluable in ensuring that a consistent set of definitions are used across the project, that ambiguities are resolved, and that synonyms and homonyms are discovered. The following information would typically be held about each data item.

Data item identifier This is a code, usually numeric, used to identify uniquely the data item. These short identifiers are particularly useful when computer tools are used since they will save typing long names, and make it easier to change the name of the data item if required.

Data item name This is the name given to the data item. It is best to be as specific as possible.

Data item definition or description A definition agreed with the users of what is meant by the data item name.

Synonyms Any synonyms that are used by differing groups of users for the data item.

Format The type of data that is to be held as the data item. Some examples are: numeric, alpha, alpha-numeric, date, pounds sterling. In the case of codes the specific format might be given, e.g. AAA99 representing three alphabetical characters followed by two digits.

Range The possible range of values that the data item could take.

Validation criteria Ways in which the values of the data can be checked for errors. Format and range serve as an initial check since if data is entered in the wrong format or out of the specified range it is clearly in error. Validation criteria is an opportunity to specify more complex checks that the data values can be subjected to. For instance these could be based on using check digits or interdependency with other data items.

Size The number of characters or bytes occupied by the data item. It is useful to include a maximum size and an average size for variable length fields since many database management systems can compact these to the space actually occupied.

Relationship Descriptions

It is advisable to describe every relationship described on the Logical Data Structure. These are not formally required by SSADM standards but in our view investigating and describing the relationships in the early stages of SSADM is an important activity because:

* relationships are often identified to fit a particular requirement, drawn on the diagram and then some time later that requirement and discussion will be forgotten and repeated;
* it helps users and analysts to understand and agree the diagram;
* often there is often more information about the relationship than could be recorded on the diagram, for instance with optional or exclusive relationships there will be conditions under which the optionality or exclusivity occurs (some database management systems can support these rules);
* volumetric information can be collected and held.

The following information might be held about a relationship:

Master and detail entities The name and entity identifiers of the master and detail entities participating in the relationship.

Relationship name If a name has been given to the relationship on the Logical Data Structure diagram this would be recorded.

Relationship description Any further information about why the relationship is required and the conditions under which it exists.

Volumetric information The volume of interest is the number of occurrences of the detail entity for one occurrence of the master entity. Minimum, average, and maximum values could all be recorded.

SUMMARY

* The Logical Data Structure is the means of describing the information held by the system and its interrelationships.
* The components of the Logical Data Structure are entities and relationships.
* The Logical Data Structure deals with entity and relationship types rather than their occurrences.
* Entities are things about which the system holds information.
* Each entity is made up of a number of data items, which are the smallest meaningful chunks of information in the system.
* Relationships relate one entity to another and indicate access from one entity occurrence to all the related ones.
* Only one-to-many relationships are drawn on Logical Data Structures.

- The Logical Data Structure diagram is supported by Entity Descriptions and sometimes Relationship Descriptions.
- Data items are described in more detail in the Data Catalogue.

2.2   Data Flow Diagrams

Introduction

Purpose of Data Flow Diagrams
A Data Flow Diagram is a diagrammatic representation of the information flows within a system, showing:

- how information enters and leaves the system;
- what changes the information;
- where information is stored.

Data Flow Diagrams are an important technique of systems analysis as a means of:

  Boundary definition. The diagrams clearly show the boundaries and scope of the system being represented.
  Checking the completeness of analysis. The construction of the diagrams, and their cross-comparison with the other major SSADM techniques, helps to ensure that all information flows, stores of information, and activities within the system have been considered.
- *Basis for program specification.* The Data Flow Diagrams denote the major functional areas of the system, and therefore the programs or program suites required.

Data Flow Diagrams in SSADM
Data Flow Diagrams may be used to represent a physical system or a logical abstraction of a system.

In SSADM, four sets of Data Flow Diagrams are developed:

- *Current Physical.* The current system is modelled in its present implementation.
- *Logical.* The purely logical representation of the current system is extracted from the current physical Data Flow Diagrams.
- *Business System Options.* Several proposed designs are developed, each satisfying the requirements of the new system. Each of these is expressed as an overview Data Flow Diagram, known as a Business System Option.
- *Required.* Using the selected Business System Option and the Logical Data Flow Diagrams, a full set of Data Flow Diagrams representing the new system is developed.

The relationship between the different sets of Data Flow Diagrams is represented in Fig. 2.27.

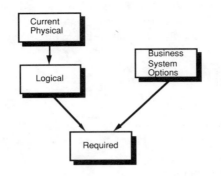

Fig. 2.27

Data Flow Diagrams are developed during the analysis stages. In design, Process Outlines, not Data Flow Diagrams, are used to define the detail of the processing. In the logical and physical design stages of SSADM (stages 4, 5 and 6) the Data Flow Diagrams are used principally to show how the total system fits together.

Advantages of Data Flow Diagrams
Data Flow Diagrams are a pictorial but non-technical representation of a system, so can be used by both technical and non-technical staff. They are used in discussions between analysts and users, and are easy to draw or amend in addition to being easily understood and verified.

One of the great strengths of Data Flow Diagrams is that it is possible to describe a system at several levels. A top-level Data Flow Diagram may show a complete system in very little detail. Each process can be separately decomposed to show the detail within. These processes can be subsequently decomposed and so on until the desired level of detail has been reached. Each level is useful to the analyst:

- the top-level diagram shows clearly the overall system boundaries, and the interfaces to other systems and users of the system;
- the detail of each individual area may be investigated in relative isolation from the rest of the system

The management of staff within a project is helped by the segmentation of the system as individuals or teams may be assigned a single top-level Data Flow Diagram process to investigate knowing that work is not being duplicated. It is a relatively simple matter to recombine work undertaken in this way.

Components of Data Flow Diagrams

External source/recipients(external entities)
An external source/recipient, often referred to as an external entity, is whatever or whoever donates information to or receives information from the system. All information represented within a system must have been obtained initially from an external source/recipient. An external source/recipient is represented on a Data Flow Diagram as an oval (or lozenge) containing the name and an identifier. The convention is that the

identifier is a lower-case letter, as shown in Fig. 2.28. An external source/recipient may be a user of the system, an external organization, a computer system, or any other source or recipient of information.

Fig. 2.28

Process

A process transforms or manipulates data within the system. Processes are represented by rectangles on a Data Flow Diagram. Each process box contains the name of the process, an identifier, and possibly a location:

- The process name is an imperative statement: 'do this' or 'do that'. It describes the processing performed on the data received by the process. For example, a process may be named 'Register new customer', but may not be named 'Manager' or 'Registration Section'.
- Process identifiers are numerical.
- In the current physical Data Flow Diagram, the location of the process is placed at the top of the box. This might be a physical location, but is more often used to denote the staff responsible for performing the process. The Logical and Required Data Flow Diagrams do not show the locations of the different processes.

Figure 2.29 shows where each of these elements are placed in a process box.

Fig. 2.29

Data store

A data store is where information is held for a time within the system. A data store is represented on a Data Flow Diagram by an open-ended box as shown in Fig. 2.30.

D1	CUSTOMER DETAILS

Fig. 2.30

In the Current Physical Data Flow Diagrams, the data stores represent real-world stores of information such as computer files, card indexes, ledgers, etc. If the Data Flow Diagrams contain both computer-based data stores and clerical data stores, it may be helpful to distinguish them by their identifiers, e.g. using 'M' for the clerical data stores and 'C' for the computer data stores. This is to avoid confusion when similarly named files are stored both clerically and on the computer system.

In the Logical and Required Data Flow Diagrams, each main data store corresponds to a set of entities from the Logical Data Structure where the stored information is described in more detail. All the data store identifiers are prefixed with a capital D.

Data flow

A data flow represents a package of information flowing between objects on the Data Flow Diagram. A data flow is represented by a line and an arrow to denote the direction of the flow of information. It is labelled with the name or details of the information represented by the data flow. This is illustrated in Fig. 2.31.

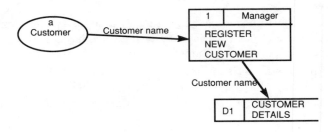

Fig. 2.31

Points to note about data flows are:

- Information always flows to or from a process. The other end of the flow may be an external source/recipient, data store, or another process.
- In the current physical Data Flow Diagrams, the data flows represent real-world flows of information. For example, these could be forms sent from one part of the system to another or telephone conversations between someone within the system and a customer.
- In the Logical and Required Data Flow Diagrams, these flows represent the data items required by a process or output from a process.

Construction of Data Flow Diagrams

The top-level Logical Data Flow Diagram of a simple banking system is shown in Fig. 2.32. Here, the main activities are the registration of new customers, the recording of deposits or withdrawals, and the closing of accounts. New customers are registered and accounts are closed by the bank manager, represented here as an external source/recipient. When an account is closed, the customer is notified by the system. Cash deposits into the account are made by the customer, and salary cheques are paid in by the employer. The bank clerk performs a balance check before allowing a withdrawal by the customer at the bank counter.

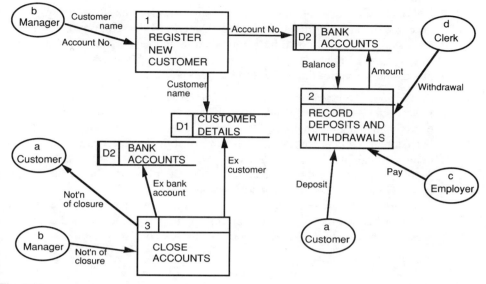

Fig. 2.32

Some general principles about Data Flow Diagrams are demonstrated by Fig. 2.32.

External source/recipients

For the Logical and Required Data Flow Diagrams only, the external source/recipient is the last to be able to alter the information. For example, when a customer makes a deposit at the bank counter, the clerk simply inputs information received from the customer without changing it in any way, so it is the customer that is shown as an external source/recipient. For withdrawals, the clerk validates the request for the withdrawal, possibly changing it before input, so it is the clerk that is shown as the external source/recipient. Similarly, for outgoing flows, the external source/recipient is the first that actually uses the information. The customer is notified of the closure of his or her account. Although the notification letter may be put into the post by a bank clerk, it is the customer that is the external source/recipient as he or she is the first person to use the information.

In the Current Physical Data Flow Diagrams, the users of the system are not separated from the processes they perform, so the external source/recipients are whoever information is sent to or received from external to the system/organization. Figure 2.31 is the Current Physical version of Process 1 in Fig. 2.32. In the current system, the manager is performing the Process 'Register New Customer', so is shown in the location area of the process box. The external source/recipient is the customer. The logical view of this is that the manager is the external source/recipient, as shown in Fig. 2.32.

Process numbering

Although the processes are numbered sequentially, this does not imply that they are executed in any particular sequence. Data Flow Diagrams do not imply sequence. Processes 1, 2, and 3 could be renumbered in another sequence and remain meaningful. Even where a process-to-process data flow exists, this need not imply that the second process must wait for the first to end before it is able to begin.

Duplication of data stores and external source/recipients

It has been necessary to duplicate certain external source/recipients and data stores to avoid overcomplicating the diagram with crossing lines. To denote that a particular data store has been duplicated in the diagram, an extra vertical line is placed at the left side of the box, as demonstrated in Fig. 2.32 by Data Store D2, Bank Accounts. Duplicated external source/recipients could be denoted by an oblique bar to one side of the oval, although this is not normal practice and is not demonstrated here.

Where objects are duplicated, it is easy to make mistakes in rewriting the names of the objects wherever they occur. Therefore, it is important to ensure that the identifiers are present to be able to reconcile different occurrences of the same objects.

Topology of the diagram

To make the diagram more readable, the external source/recipients have been arranged around the edges of the diagram, and the data stores placed towards the centre of the diagram. This is good practice rather than a particular rule. For clarity, no more than 12 processes should be shown on a single Data Flow Diagram. It is important to remember that one of the the main purposes of a diagram is to act as a means of communication, so legibility and clarity are as important as the technical content of a diagram.

Levels of Data Flow Diagrams

Each process on a Data Flow Diagram may be broken down into several processes which are shown on another Data Flow Diagram. This is described as decomposing the Data Flow Diagrams. The Data Flow Diagram which is a result of this decomposition is one level below the Data Flow Diagram containing the original process.

The Data Flow Diagram that describes the entire system within a single diagram is the 'top-level' or 'level 1' Data Flow Diagram. The Data Flow Diagrams that are expansions of processes at the top level are 'level 2' Data Flow Diagrams (see Fig. 2.33). Levels below this are called 'level 3', 'level 4', etc. Processes that are not further decomposed are 'bottom-level' processes. Processes from the top-level Data Flow Diagram may be broken down to a number of levels if they are complex or may not be broken down at all if they are relatively simple. Thus, it is possible to have bottom-level processes appearing at all levels of the Data Flow Diagrams. Although not formally documented as part of SSADM, it is useful to denote which are bottom-level processes on the Data Flow Diagrams.

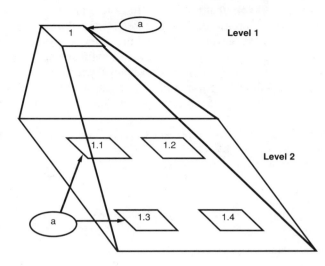

Fig. 2.33

If a process is decomposed, the identifiers of the lower-level processes are prefixed by the identifier of the higher-level process. For example, if Process 5 is decomposed, the lower-level processes will be identified as 5.1, 5.2, etc. Similarly, if Process 5.1 is subsequently decomposed, the lower-level processes will be 5.1.1, 5.1.2, and so on.

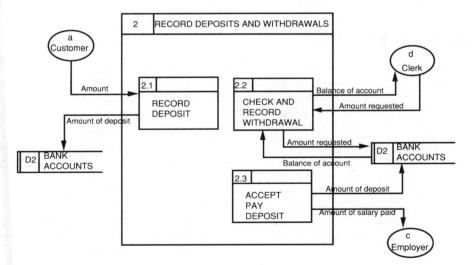

Fig 2.34

The second-level Data Flow Diagram of Process 3 is shown in Fig. 2.34. The frame surrounding the lower-level Data Flow Diagram denotes the boundary of the higher-level process. The identifier of the higher-level process and the name of the process are put at the top of the frame. Note that all of the flows to and from the higher-level box have been either duplicated or broken down into several flows at the lower level. If new data flows are identified at the lower level which cross the frame, these should be reflected at the higher level so that consistency between the levels is maintained.

So that it is clearly understood what is represented by each bottom-level process, it is useful to write a brief description of the process with an indication of when the process is triggered into action. This description is called an Elementary Function Description.

SUMMARY

Data Flow Diagrams are an important systems analysis technique for representing the flows of information within a system.

The Data Flow Diagrams produced in SSADM are:

- Current Physical;
- Logical;
- Required (based upon the selected Business System Option).

The components are:

- external source/recipients;
- processes;
- data flows;
- data stores.

Each process may be further described by:

- decomposition to another level of Data Flow Diagrams;
- Elementary Function Descriptions.

2.3   Entity Life Histories

What is an Entity Life History?

A major reason for building computer-based information systems is to provide up-to-date and accurate information. Information is constantly changing, for example the number of beds available in a hospital, the price of petrol, and people's names and addresses. An information system must be able to keep track of these changes. The previous sections have described how the system is modelled from the viewpoint of information flows (Data Flow Diagrams) and from the viewpoint of the information that is held (Logical Data Structures). Entity Life Histories model the system from the viewpoint of how information is changed. What the Entity Life Histories show is the full set of changes that can possibly occur to the information within the system, together with the context of each change.

Initially, each entity within a system is examined in isolation as this is a manageable unit of information to model. It is the stimuli of the changes that are modelled rather than the processes that operate to implement those changes. By specifying the set of changes to each entity, a composite picture is formed, eventually specifying the full set of changes that will occur within the system.

An Entity Life History is a diagrammatic representation of the life of a single entity, from its creation to its deletion. The life is expressed as the permitted sequence of events that can cause the entity to change. An event may be thought of as whatever brings a process into action to change entities, so although it is a process that changes the entity, it is the event that is the cause of the change.

An example will explain Entity Life Histories more clearly. Imagine that Maurice Moneybags has decided he wants to open a bank account at Cash & Grabbs Bank. When Maurice has persuaded Mr Cash, the manager, that he would be suitable as a customer, Mr Cash turns to his computer terminal and records Maurice's new bank account code in the system. The Logical Data Structure of the bank computer system contains an entity called Bank Account. The event occurrence that creates the Maurice Moneybags occurrence of the entity Bank Account is the opening of the account by Mr Cash. This event occurrence and the ones that follow it are:

- Account opened for M. Moneybags
- Cash Deposit £2000
- Cheque Cashed for £20
- Direct Deposit £1000
- Cheque Cashed for £20
- Direct Deposit £2000
- etc.

Percy Penniless is another customer at the bank, the event occurrences that affect his bank account are:

- Account Opened for P. Penniless
- Pay Deposit by Credit Transfer £500
- Cheque Cashed for £ 200
- Cheque Cashed for £300
- Cheque Cashed for £300

Other accounts may behave in similar ways, none of which are precisely the same. However, it is possible to build up a general picture that will fit all occurrences of Bank Accounts at the Cash & Grabbs Bank. Basically, all accounts are opened, and several deposits and several withdrawals may be made. The way that these events affect the entity Bank Account can be summarized using a diagrammatic notation, as shown in Fig. 2.35. Figure 2.35 is read from left to right, and wherever there is a structure going down the page, this is followed through before continuing along the left-to-right progression. This Entity Life History shows that the first event to affect the entity Bank Account will be Account Opened for all occurrences. Next, the Account has a life which is a series of transactions. Each Transaction is one of: a Pay Deposit, a Direct Deposit, or a Cheque Cashed. After an undefined number of Transactions have taken place, the Account will be closed and finally deleted.

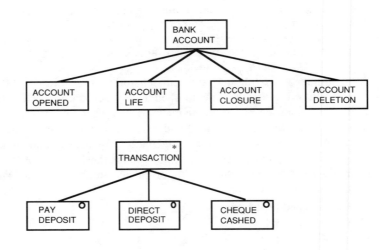

Fig. 2.35

The elements of the notation used in this example are:

- *Sequence:* the boxes read left to right from Account Opened to Account Deletion. Thus Account Closure must happen before Account Deletion.
- *Selection:* the boxes with small circles in the top right corners are alternatives for one another, so each Transaction is one of Pay Deposit, Direct Deposit or Cheque Cashed.
- *Iteration:* the box with the asterisk in the top right corner represents an iteration, many Transactions can occur one after another.

All the elements of the notation are described in detail in the next section.

Terms and notation

An entity may be affected by several different events. In the banking example the events Account Opened, Pay Deposit, Direct Deposit, Cheque Cashed, Account Closure, and Account Deletion all affect the entity Bank Account.

An event may also affect several entities. For example, when the Bank Account of Maurice Moneybags was opened, a 'Maurice Moneybags' occurrence of the entity Customer would also have been created. The event Account Opened affects two entities: Bank Account and Customer.

Thus an entity can be affected by a number of events, and an event may affect a number of different entities. To describe the particular interaction between a single event and a single entity, the term 'effect' is used. The box on the Entity Life History of Bank Account represents the effect of the event Account Opened on the entity Bank Account. Although the box represents the effect of the event the name inside the box is always the name of the event.

Within an Entity Life History diagram, the effect boxes have no other boxes directly beneath them. Intermediate boxes with other boxes directly beneath them are called 'nodes'. Nodes have no significance other than in expressing the valid event sequences within the context of the Entity Life History diagram. Within an Entity Life History, the names in the boxes will reflect the names of events if they are effect boxes and the names will reflect a particular section of the Entity Life History if they are node boxes.

An Entity Life History diagram is constructed for each entity on the Logical Data Structure. The entity is placed in a box at the top of the diagram, and all possible progressions through the life of the entity are overlaid on one another, to form a picture that fits every occurrence.

Entity Life History diagrams use the following diagramming components:

1. Sequence.
2. Selection.
3. Iteration.

All Entity Life Histories can be built up using just these three components. However, certain complex situations can be simplified by the use of two other conventions:

4. Parallel structures.
5. Quit and resume.

For clarity, the diagramming component types may not be mixed at the same level within the same part of the diagram. Thus, an iteration may not be placed at the same level as a sequence; instead, a node is placed within the sequence and the iteration dropped down a level. The component types are described here, but their use within SSADM is described in later sections.

Sequence

A sequence is represented by a series of boxes reading from left to right as shown in Fig. 2.36.

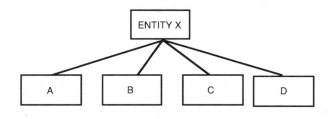

Fig. 2.36

The box labelled A will always be the first to occur, followed by B which in turn is followed by C then D. This is the only possible sequence. Although the sequence may be thought of as a progression through time, there is no indication of the time intervals between the boxes within a sequence. These could span minutes, hours, days, or years.

Selection

A selection defines a number of effects or nodes that are alternatives to one another at a particular point in the Entity Life History. A selection is represented by a set of boxes with circles in the top right corners as shown in Fig. 2.37.

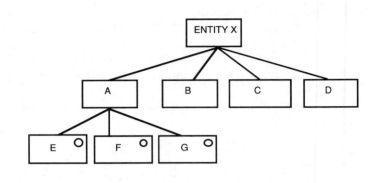

Fig. 2.37

As node A is at the beginning of the Entity Life History, this diagram shows that an occurrence of entity X must be created by only one of three events: E, F or G. If there is a requirement to show that one of the options need not be selected, a 'null' box may be added as shown in Fig. 2.38. This extension to the banking example shows that a Bank Account could start with a Pay Deposit, a Direct Deposit, or neither of these.

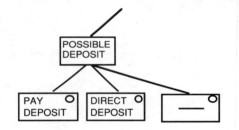

Fig. 2.38

The null box does not represent an effect or node, but is a notational device to indicate the situation where something may or may not happen. If the null box is selected, the Entity Life History continues directly to the next node or effect.

Iteration
An iteration is where an effect or node may be repeated any number of times at the same point within an Entity Life History. A restriction upon the iteration is that each occurrence of the iteration must be complete before the next begins. This is most relevant where a node is being repeated. An iteration is represented by an asterisk in the top right-hand corner of a box as shown in Fig. 2.39.

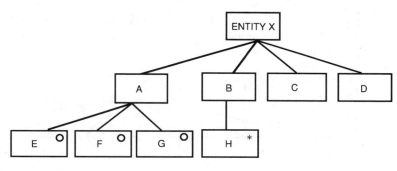

Fig. 2.39

After entity X has been created by E, F, or G, the event H may affect the entity any number of times. Here it is important to note that 'any number of times' includes none, so an iteration is another way of showing that something may occur or it may not. However, the iteration must not be used where an effect or node occurs either once or not at all. In this situation the null box is more accurate as in Fig. 2.38.

Parallel structures

A parallel structure is used in the situation where effects or nodes occur in no predictable sequence or concurrently. A parallel structure is shown as a parallel bar on the Entity Life History diagram as in Fig. 2.40. This represents the situation where the sequence of K, L, and M occurs at this point in the Entity Life History, and the event N may affect the entity a number of times during this sequence. The node I, representing the sequence, and the node J, representing the iteration, are shown under the parallel bar.

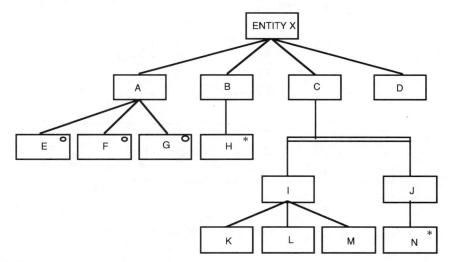

Fig. 2.40

Figure 2.40 could be drawn without the use of the parallel structure, as shown in Fig. 2.41. In order to shown that the event N may occur anywhere in the sequence of K, L, and M, the iteration of N must be repeated several times. The diagram is made much more clear by the use of the parallel structure.

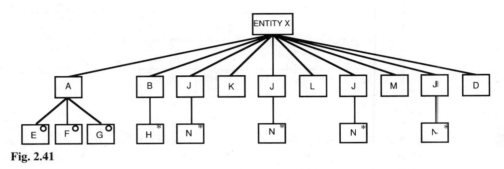

Fig. 2.41

Quit and resume

The quit and resume convention is used in situations where the diagramming conventions excessively constrain the Entity Life History, or force a very complex artificial structure in an attempt to model a particular situation. The use of this convention allows a quit from one part of an Entity Life History diagram to resume in another part of the diagram. In the simple case, a Q and a number are placed by the right-hand side of a box on the diagram, and a corresponding R and the same number are placed by the left-hand side of a box in another part of the diagram. This shows that the effect or node that follows in sequence the box annotated with the Q is always the box annotated with the corresponding R. (See Fig. 2.42 for an example, where Q1 and R1 are the quit and resume, respectively.)

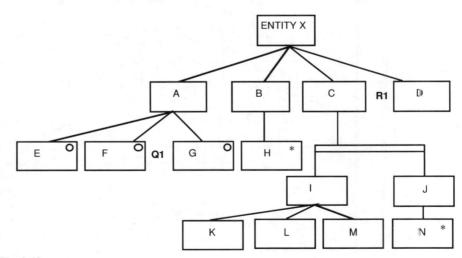

Fig. 2.42

When an occurrence of entity X has been created by event F, the next event to affect the entity will always be D and cannot be H, K, or N.

To avoid any ambiguity, Fig. 2.42 could be redrawn to give the structure shown in Fig. 2.43.

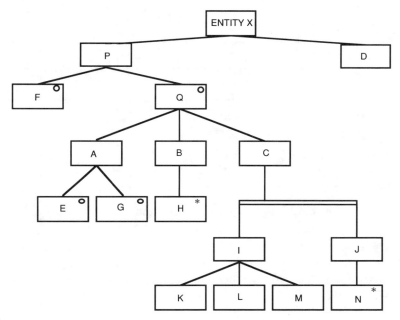

Fig. 2.43

Two new nodes, P and Q, have been introduced into the diagram to ensure that objects of the same type remain at the same level. In practice, if quits and resumes are all removed, the diagrams become very complex, reducing the clarity of the diagrams. Thus a compromise between ambiguity and clarity is made.

In the banking example, it is possible to reopen a bank account that has been closed, but not yet deleted. In this case, the event Account Reopened causes a quit back to the part of the Entity Life History before Account Closure. This backwards quit is shown by Q1 and R1 in Fig. 2.44.

Another use of the quit and resume convention allows a quit from the main structure of an Entity Life History, and a resume at a box on a stand-alone structure. This is used where an event might occur at any time, altering the sequence of the Entity Life History. As it is impossible to predict where on the diagram the quit might occur, no Q is placed on the structure. Instead, a sentence indicating the area of the Entity Life History the quit might occur from and the circumstances that cause the quit is placed at the bottom of the diagram. The stand-alone box or substructure is annotated with the R that corresponds to the Q detailed in this sentence. An example of this random quit is shown by Q2 and R2 in Fig. 2.44. Here, the death of a customer may occur at any time after the account has been opened. If a customer dies, the normal sequence is no longer applicable and this account is placed into suspension. The account is deleted when the customer is finally deleted from the system.

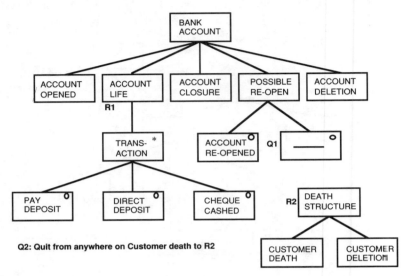

Q2: Quit from anywhere on Customer death to R2

Fig. 2.44

If necessary, it is permitted to quit from this substructure back into the body of the Entity Life History.

In general, it is possible to have more than one quit point with the same identifier on the same diagram, but there must only be one resume point with the same identifier to avoid ambiguity.

SUMMARY

- Entity Life Histories model the changes to the information held in a system.
- Events that cause the system to update entities are reflected within Entity Life Histories.
- The particular set of changes in an entity caused by an event is an effect.
- An Entity Life History diagram shows the sequences, selections, and iterations of the effects of events on a single entity.
- Two additional conventions—parallel structures and quit and resume—are used to simplify the diagrams.

2.4   Relationship between the three views

Overview

It is a very important principle of SSADM that the different views of the system should be closely related to one another. Each view is built up separately and is then checked for consistency and completeness by comparison with the other views. The required system Data Flow Diagrams, Logical Data Structure, and Entity Life Histories are related to one another in defined ways and can be checked against one another before being signed off as correct.

The degree to which the techniques can be checked against one another depends upon the SSADM stage. The rules for cross-checking the Logical Data Structure with the Data Flow Diagrams cannot be strictly applied for the current system as they will reflect the current situation which may not be very logical. Thus, the relationships described in the following sections concentrate upon the rules that apply in the required system definition with some indication of how they can be used in earlier stages.

Relationship between Data Flow Diagrams and Logical Data Structures

Logical Data Structures reflect the structure of stored data. A Data Flow Diagram shows data moving about the system and being stored in data stores. All data items should appear in an Entity Description if they are held within the system and will probably be shown flowing around the system on data flows. The rules that govern the relationship between the two views are described here.

Each data store should represent a whole number of entities.

As both entities and data stores represent stored data, there is a very close relationship between them. As an entity is a grouping of related data items, it would not be logical that it could be split across more than one data store. Also, as the Logical Data Structure is the more detailed view of data, it is unlikely that a data store would represent a smaller grouping than a single entity. These principles have been translated into the following rules:

- A data store is related to one or more entities.
- An entity may not appear in more than one data store.

The only exceptions to these rules are:

- In the current system, data is often duplicated, so it is possible for an entity to appear in more than one data store.
- Transient data stores do not represent stored data and will therefore not be related to any entities on the Logical Data Structure.

The precise relationship between particular entities and data stores is formally documented using an Entity/Data Store Cross Reference form. This form is constantly referred to when checking either technique for consistency.

Data items on data flows should belong to entities.

Another relationship between Data Flow Diagrams and the Logical Data Structure is in relating the data items flowing around a system to the data items attached to entities. This can be used when checking the data flows into and out of data stores to ensure that the data items denoted on the data flows appear in the Entity Descriptions relating to the data store.

The only data items that may not appear on a Logical Data Structure are:

- Transient items such as partial results.
- Derived items for output from the system.

Also, some data flows may just have a label (such as 'Errors') which will not have an immediate relationship to the Logical Data Structure view.

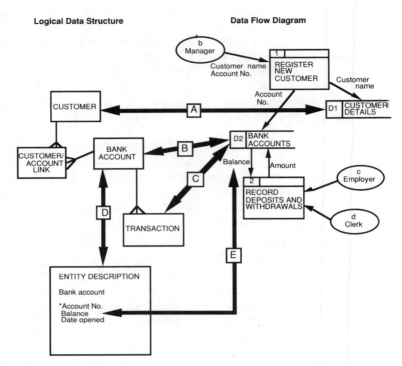

Fig. 2.45

The relationships between the Data Flow Diagrams and Logical Data Structure are shown in Fig. 2.45, which shows how some areas of the banking example are related:

A. The Customer and Customer/Account Link entities relate to the Customer Details data store.

B. The Bank Account entity relates to the Bank Accounts data store.

C. The Transaction entity relates to the Bank Accounts data store.

D. The Bank Account entity is described by a set of data items.

E. The Balance data item on the Entity Description relates to the Balance data item on the data flow.

Relationship between Logical Data Structures and Entity Life Histories

An Entity Life History is constructed for each entity (other than operational masters) on the Required Logical Data Structure (and later updated to reflect changes in the Composite Logical Design). The starting point for Entity Life Histories is the ELH Matrix in which all entities from the Logical Data Structure are listed down one side of the matrix and events added to the other side. The Entity Description and relationships to other entities are examined to ensure the completeness of the Entity Life History:

- Each data item must be created, (possibly amended), and deleted by events on the Entity Life History.

Changes in relationships, e.g. change of master entity, may be events that should be taken account of.

The Entity Life Histories of entities may be affected by the Entity Life Histories of entities related to them as masters or details.

Thus, the construction of Entity Life Histories is done by referring constantly to the Logical Data Structure and the supporting Entity Descriptions.

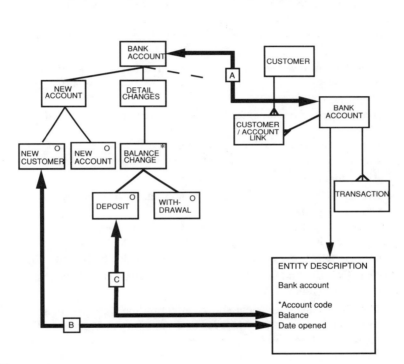

Fig. 2.46

Figure 2.46 illustrates the relationship between an Entity Life History and the Logical Data Structure. The relationships between the two diagrams that are illustrated in Fig. 2.46 are:

A. An Entity Life History is drawn for each entity on the Logical Data Structure. Here, the Bank Account Entity Life History is for the Bank Account entity on the Logical Data Structure.

B. In creating the entity, the events New Customer and New Account will set a value for the data item Date Opened. This data item will not be updated by any subsequent event.

C. The Deposit and Withdrawal events will update the Balance data item each time they affect the Bank Account entity.

Relationship between Data Flow Diagrams and Entity Life Histories

The events reflected in the Entity Life Histories are triggers to the Data Flow Diagram processes that update the system data. Therefore, there should be some indication of the events on the Data Flow Diagrams. If the event is the arrival of input data, then events can be directly related to input data flows. If the events are not associated with particular data flows (as for time-based events), the relationship between events and the Data Flow Diagrams is not so clear cut. However, each process that updates a main data store (i.e. one or more entities on the Logical Data Structure) must have an event (or events) associated with it.

To summarize, the relationships between Entity Life Histories and Data Flow Diagrams are:

- Each bottom-level process on the Required System Data Flow Diagram that updates a main data store will be triggered by one or more events that appear on the Entity Life Histories.
- If an event is from an external source, it will probably be associated with an input data flow.
- If a process is triggered by something other than the arrival of data, the event will not be shown explicitly on the Data Flow Diagram but can be inferred by inspection

The way that this works in practice is:

- Events are initially identified by looking at the Data Flow Diagrams and documenting what triggers the processes.
- After Entity Life Histories have been completed, any new events that have been identified are reconciled with the Data Flow Diagrams. If events are identified that do not map to processes, new processes will be added to the Data Flow Diagrams.

Figure 2.47 illustrates the relationship between Data Flow Diagrams and Entity Life Histories. The relationships shown are:

A. The New Customer event shown affecting the Bank Account entity is represented by the data flow from the Manager external entity to Process 1. This was identified as an event affecting this entity because there is a data flow from the process to the Bank Accounts data store, showing that an entity is being updated.

B. The Deposit and Withdrawal events are related to the inputs to Process 2. On closer inspection, several different events may be identified by comparing the Entity Life Histories and Data Flow Diagrams as the input from the Employer will be a Pay Deposit and the input from the Clerk will be a Cash Deposit (see Sec. 2.3).

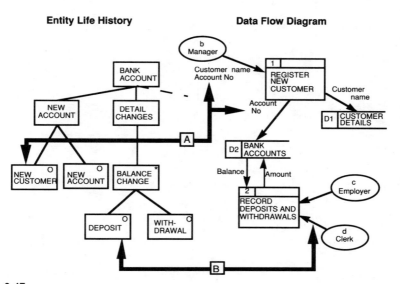

Fig. 2.47

SUMMARY

Data Flow Diagrams, the Logical Data Structure, and Entity Life Histories are all related to each other in some way. It is important that the diagrams are made consistent at all times so that at any stage in a development project the three views are showing different aspects of the same thing.

This is a very important aspect of SSADM. At each stage, each diagram can be used to check the consistency and completeness of the others. If a software support tool is to provide adequate support for SSADM, it is important that these relationships are reflected in the underlying structure so that the relationships described here can be checked automatically.

Part 2 Working through SSADM

Part 2 follows the structure of SSADM with a separate chapter on each of the six stages of SSADM. Each chapter begins with an introductory section describing the stage. The remainder of each chapter is divided into several sections each describing a step.

The overall objective of Part 2 is to show how SSADM might be used on a typical project. The reader is taken through the development of the major SSADM end-products and shown how they should be developed.

To illustrate the development of a SSADM project we have used a case study. This concerns the activities of a fictional vehicle rental company, Yorkies. A description of the current system is given in Appendix C.

Exercises are given at the end of many of the sections. These test the techniques used in the step described in the section. Suggested answers are given in Appendix D.

3. Analysis of system operations and current problems (stage 1)

3.1 Introduction

Objectives of the stage

When a computer system is introduced into an organization, in most cases it is to support the work already done by that organization, albeit with modifications or enhancements. Although the workings of the future system may differ substantially from those of the current system, the information held and the major functions often remain relatively unchanged. Thus analysis of the existing system provides a firm basis for the design of the future system.

Additional reasons for performing a current system investigation are:

* *To understand the scope of the project*. The complexity of the system can be determined, enabling the rough estimates of effort required to complete the project to be determined. This helps to plan and resource the project.
* *To increase confidence*. The current system is known in great detail by the users who may not have any idea of what a computer system can give them. It is the development team's task to specify the job that must be done and how that job can best be supported by a computer system. Documenting the current system is a means of reassuring the users that the analysts understand the nature of the problem fully and are competent to carry the work forward into the design of the required system. The analysts become confident that they understand the business of the system.

In order to be able to design and justify the new system, several aspects of the system are investigated:

* *Operations and data*. In understanding the current operations and data, the requirements are more easily understood.
* *Problems inherent in the current system*. Examining the problems encountered in the current system ensures that they will not be replicate d in the new system.
* *Boundaries of the system*. It is important, throughout the project, to attempt to draw accurate boundaries around the area of investigation to avoid needless effort on areas that lie outside the boundaries or, more importantly, to ensure that all of the areas inside the boundaries are investigated. The boundary of the current system investigation should be explicitly stated and agreed with all those concerned.

- *Costs and volumes*. These act as a basis for system justification and sizing in the analysis of requirements.

It is important to note that an investigation of the current system does not constrain the project team to reimplement what is present in the current system. In the second stage of SSADM, the Business System Options step takes a fresh approach to meeting the objectives of the system. This may lead to a complete restructuring of processes and data and a new system that is nothing like the old.

The current system

The different types of current system may be classified under the following headings.

Fully clerical (Fig. 3.1)

Fig. 3.1

The new system might be required to either:

- Replace a system currently using manual procedures and paper files. An example of this might be a personnel system which is implemented entirely on card indexes. The new computer system would be required to store the records on computer media.

Or

- Monitor a system that will remain in place. An example of this might be a system that monitors stock in a warehouse. The stock will continue to be bought and distributed in the same way, but the new system would be required to monitor the levels of stock and to control reordering of stock.

Although the requirements of these two types of system will be different, the investigation of their current systems follow the same pattern. All the information currently held or currently monitored and all the procedures are modelled using the SSADM techniques.

Partly computerized (Fig. 3.2)

Fig. 3.2

The new system may be required to replace an existing computer system because it:
- does not meet the user's requirements;
- has reached the end of its maintained life;
- requires major enhancements.

An example of this might be where a payroll system is already computerized, and the corresponding personnel system is clerical. A new requirement to computerize the personnel records might prompt the requirement for a fully integrated personnel and payroll system, even though the payroll is already implemented as a computer system.

The computer system forms only a part of the current system modelled by SSADM. The clerical procedures surrounding the system will also form a part of the investigation, although the precise boundaries will be set by the user.

No apparent current system (Fig. 3.3)

Fig. 3.3

Sometimes, a completely new requirement will arise, for example when new legislation is introduced or an organization moves into a new business area. It may appear, in these cases, that there is no current system to investigate. However, if a similar system already exists within the organization, it might be useful to perform a brief investigation on this system to establish areas of commonalty of both information and procedures.

An example of this is where an organization that is currently archiving and retrieving paper records decides it has a need for a similar system to archive and retrieve files stored on computer media. Although the requirement has no real current system, the system for archiving and retrieving paper files might provide a useful basis for the investigation.

In cases where there is no similar system within the organization it may be necessary to look at how other organisations deal with similar problems. If the requirement arises from a new law then it will be necessary to analyse the requirement expressed within the legislation and supporting documentation.

Often it appears that there is no existing system although the activities are currently performed. This may indicate that the current system is a highly informal one (see Land and Kennedy-McGregor in Galliers, 1987). All systems have both a formal and an informal component. The formal component operates through rigid rules and uses highly structured information such as computer files and manual forms. The informal component operates through, intuition, value judgements, and unstructured information such as conversations and prose. If the objective is to develop a computer system to support a largely informal system then analysis and formalization of the current informal system is essential. Stage 1 of SSADM in this case would require analysis of the current informal system as well as the formal one.

Overview of stage 1

Throughout stage 1 of SSADM, the development team must find out about the system by a number of fact-finding techniques:

- interviewing;
- studying the current system's documentation;
- circulating questionnaires;
- observing the system in practice;
- looking at the results of previous studies;
- conducting surveys.

These are well-known techniques described in many textbooks (e.g. Daniels and Yeates, 1988) and will not be described here. Certainly, whatever other techniques are chosen, it will be necessary to interview one or more users in order to understand their perspective and to appreciate the problems associated with the current operations.

The information collected is expressed as: a set of detailed Data Flow Diagrams representing the current system, a Logical Data Structure of the current system's data, and the initial Problems/Requirements List for the project. In addition, information about the project, especially volumetrics and costs, is gathered during the analysis of the current system. This is used later for cost justification and sizing of the new system.

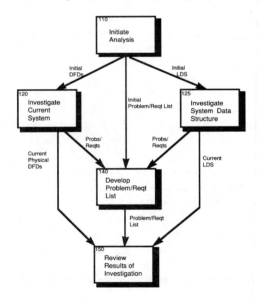

Fig. 3.4

Structure of stage 1

Figure 3.4 shows the five steps of stage 1, and their relationship to one another.

After the project is initiated in step 110, the current system is documented by the use of Data Flow Diagrams (step 120) and a Logical Data Structure (step 125). A Problems/Requirements List, initiated in step 110, is expanded within step 140. The results of the current system investigation are reviewed in step 150 before proceeding to stage 2.

3.2   Initiate analysis (step 110)

Overview of the step

This first step of the project may be thought of as two separate, but related, parts:

- General project set-up.
- Initiation of analysis.

The objective of this step is to provide a firm foundation for the whole of the development with a detailed framework for the first three stages of the method. The scope and size of the project are agreed and plans are drawn up for the first three stages of the project. Much of the activity described in this section belongs under the heading project management and control. Some further reading on this topic is suggested in the bibliography.

General project set-up

The tasks to be performed will depend upon the depth of any previous studies. For instance, the users may have independently produced a detailed description of their requirements or the application may have been the subject of a feasibility study or part of a strategic study. In these cases, the development team will study the documentation available in order to proceed in the way desired by the users. When there is little information available, the development team will need to spend more time in talking to the users and sorting out their terms of reference. In all cases, the following documents are produced.

Terms of reference

A short document known as the terms of reference should be drawn up and agreed between the senior users (usually the people who requested the system and who have the financial authority to pay for it) and the development team. If a strategy or feasibility study has been done, then this could be a very detailed document, giving clear boundaries for the project, and precise budgets and timescales. The recommendations of the previous study would form the basis of the terms of reference for the next study. If no previous studies have been performed then the terms of reference might be rather vague statements about scope, budgets, and timescales.

Terms of reference

1. To design a computer system to support the vehicle rental, driver administration, customer records, and invoicing areas of Yorkies Ltd.

2. To investigate ways of improving the efficiency of the operations of the company in the areas specified in 1.

3. To investigate extending the system to include the administration of one-way hires and the acceptance of non-registered customers.

4. The system will not be required to replace existing staff but the increased efficiency will be expected to increase the income from customers with less wastage of resources in the use of agency drivers.

5. The team of a contract senior systems analyst and a junior analyst are required to use SSADM for the analysis and design of the system. They will report back to the Yorkies finance director. The team should report back with their proposals for a new system (stage 3 of SSADM) within two months of starting the study.

Fig. 3.5

It is important that some sort of contract should be agreed between the u ers and the developers. This is often the first formal statement by the users of what they want and thus forms the basis for the subsequent work.

As an example of the type of document that might be agreed, the terms of reference for the Yorkies Ltd project are listed in Fig. 3.5. In this case no previous studies had been performed.

Definition of boundaries and scope of investigation

It is important, from the outset of a project, to be as clear as possible about the boundaries and scope of the investigation. This ensures that only the areas of interest are studied and that nothing is left out. It is often the case that a much larger proportion of the system is investigated than is necessary, extending the timescales of the project needlessly and wasting time and resources. Missing areas out can have similar effects as much effort will be wasted later in the project trying to cover the areas missed in earlier stages.

The techniques of Data Flow Diagrams and Logical Data Structures help in the definition of boundaries and scope:

* Data Flow Diagrams show data flowing across the boundaries of the system from processes within the system to external sources and recipients, and vice versa. This shows clearly where the boundaries of the investigation will lie.
* Logical Data Structures show the objects, or entities, within the system that will be investigated.

In large systems, a single user with detailed knowledge of every part of the system is difficult to find. If this is the case, an overall picture of the current system might be built up from a set of partial views obtained from a range of users. Data Flow Diagrams and a Logical Data Structure of each partial view are easy to combine to give the complete picture. Alternatively, a single Data Flow Diagram and Logical Data structure might be drawn in consultation with a group of users together.

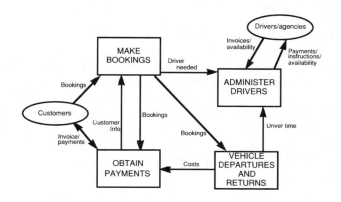

Fig. 3.6

The overview Data Flow Diagram produced for the Yorkies Ltd system from these initial discussions with users from both the local offices and the head office is shown in Fig. 3.6. Only the processes and a few external source/recipients are shown, but it does give a fairly clear picture of the scope and boundary of the system under investigation. The supporting overview Logical Data Structure is shown in Fig. 3.7.

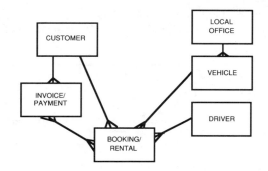

Fig. 3.7

The conventions of Logical Data Structuring have not been adhered to rigidly as the purpose of Figs 3.6 and 3.7 is to give an outline of the task. The full Data Flow Diagrams and Logical Data structure of the current system are constructed in later steps as described in Secs 3.3 and 3.4.

The project plan
With this first idea of the scope of the project it is possible to formulate estimates for effort requirements. As a rule of thumb, the number of processes on the overview Data Flow Diagram, the assessed complexity of the project, and various other factors can be used to base estimates of how long each step will take to complete. For example, each process on the overview Data Flow Diagram can be assumed to be expanded to a further four processes at the second level. An estimate of the man days required to complete the full set

of Data Flow Diagrams can be built up by multiplying the number of processes by the days required to complete one process.

Where this mechanistic approach is taken, a spreadsheet software package could be used to estimate readily the total effort required for the project. Additional tasks performed in the project, such as meetings or preparation of reports, would be added in to this total, and an idea of the elapsed time for the project may be derived. Any constraints on the time available for the project can be scrutinized and compared to the estimates. This might lead to a change in the terms of reference if there is a significant disparity.

Any estimates derived at such an early stage of the project are unlikely to be very accurate so it is important that the project management are prepared to revise the plans as more of the analysis work is completed.

Selecting the development team

One of the basic principles of SSADM is that of user involvement. The term 'user' covers a range of roles: directors or sponsors, managers, and operators. Each of these roles has a particular part to play in the project. User directors will make policy decisions associated with the project, sanction the expenditure on the system, and release their staff for development work. The total commitment of the user directors is a prerequisite for the success of the project. User managers and operators will be interviewed, participate in quality assurance, and may be be seconded to the development team.

It is necessary that a group of users who have the authority to make decisions on all aspects of the development project is identified and appointed. The organization chart for Yorkies is shown in Fig. 3.8, indicating the user areas involved in the project.

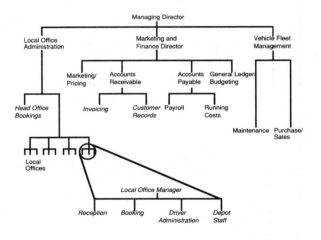

Fig. 3.8

A cross-section of the organization should be selected: if the users are at too low a level, they will not have the authority to make major decisions, and if they are all at too high a level, they will be more out of touch with what the future 'hands-on' users of the system will require. It is important that these users are committed to the project, both in terms of a positive interest in the application and in terms of time made available to the analysts for interviews and review meetings. If no commitment is made by the users at this stage, the risk of the project failing is increased significantly. This group of users will attend quality assurance reviews. The composition of the team should not change through the course of the project to ensure continuity.

For the project team, the project manager and teams of analysts should be clear as to their responsibilities. It works well in practice to have a mix of users and systems analysts working in teams. Although able to use the basic techniques of SSADM, inexperienced analysts and users will need to have a more experienced analyst to refer to, as SSADM does not replace traditional systems analysis skills.

Project management and control procedures
SSADM does not include procedures for project management and control. Before the project starts standards for project management and control should be adopted. There are many such standards which generally encompass the following: what is to be produced in each stage of the project; that suggest management and user structures for the project; that define how estimates, plans, and budgets are developed, maintained, and presented; and that define how quality assurance should be performed. Many organizations have set such standards or adopted a commercially available project management and control methodology. A good example of the latter is PROMPT (this is described further in Yeates, 1986), which has become a standard for UK government use and has been adapted to complement SSADM. The quality assurance procedures described in this book are based upon the PROMPT approach.

The use of software support tools
If no software tools to support SSADM are already available to the project team, then the possible procurement of such a tool should be decided upon at this step. Experience has shown that productivity can be increased significantly when the right tool is chosen to produce the SSADM documentation. Also, the use of such tools has an impact on the subsequent maintenance of systems developed using SSADM.

Technical advice and training
Advice might be required if the development team are inexperienced in the use of SSADM. This is not because the method requires a high degree of expertise, but experience helps in the use of the techniques and procedures. The role of a technical consultant should be clearly defined and might involve:
- advice on the use of techniques;
- presentations to managers and users;
- estimation of resource requirements;
- tailoring of the method to particular projects;
- quality assurance of work produced on the project.

Training will be required for any staff using SSADM for the first time. Typically, commercial SSADM courses are split into two one-week courses for experienced system analysts. The first week concentrates on stages 1, 2, and 3, and the second week on the remaining stages. Inexperienced analysts are generally advised to go a four-week course which covers basic system analysis skills as well as SSADM. Training courses for users and for project managers may also be appropriate.

Initiation of analysis

Once all of the parameters of a project are set, the development team is able to prepare for the investigation of the current system. The tasks and products required here are:

Detailed allocation of project team members to tasks Although an overall project plan has been drawn up, it remains to detail the actual tasks to be performed by the analysts, and to allocate individuals or teams to tasks.

Interview plans It is important to interview a good cross-section of users, at all levels of the organization, to be able to build a complete picture of the system. A plan should be drawn up for agreement with the user management, detailing the interviewers and interviewees for each interview, and giving a timetable for the interviews.

Initial Problems/Requirements List In this step, there will be some initial discussions with users and their problems with the system will begin to emerge. Their comments may take the form of 'The way this works is a real nuisance...' or 'If only I could do this...'. These comments should be documented on an initial Problems/Requirements List which will be fully developed as the series of interviews progresses. (A full description of the development of the Problems/Requirements List is given in Sec. 3.5). If comments about the users' problems were to be documented only as part of a discussion record for the whole interview, they are likely to be missed at a later stage.

Authorization to proceed

As the products produced in this step are so critical to the success of the project they are normally the subject of a quality assurance review. This should result in a formal agreement that is made with the senior users to proceed with the project. This establishes the baseline for the first three stages of SSADM.

SUMMARY

- Step 110, *initiate analysis,* involves two main activities: general project set-up and the initiation of stage 1.
- As this step sets up the framework for the whole project it is critical for its success.
- The activities required for project set-up will depend on what has gone before but will include some of the following:

  —definition of boundaries and scope;
  —project estimating and planning ;
  —definition of terms of reference;
  —setting up a development team;
  —defining project control procedure;
  —selecting software support tools;
  —arranging technical advice and consultancy.
- The initiation of analysis will involve:

  —allocation of the development team to specific tasks;
  —development of interview plans.
- The various plans developed are subject to a formal quality assurance review which should result in authorization for the project.

3.3   Investigate current system (step 120)

Objective

The objective of this step is to represent the workings of the current system using Data Flow Diagrams and to agree these with the users.

Approach

The Data Flow Diagram is a meeting-point of the analysts and users. The use of a diagrammatic model gives a precise and concise representation of the current system that can be used as the basis for further analysis and design. The diagrams use the language of the current system so that, for example, a 'Pink form 13A' that is used will appear on the Data Flow Diagrams as a 'Pink Form 13A'.

The development of the Data Flow Diagrams is done jointly by the users of the system and the analysts. Typically, the Data Flow Diagrams would be started by the analysts after listening to a description of the current system given by its users. Then would follow a number of further discussions in which the users would work closely with the analysts to refine the diagrams until they represent the current system accurately.

As the users are so closely involved with the development of the Data Flow Diagrams, it follows that they must understand the notation and their purpose. This may require some training.

Starting the Data Flow Diagrams

With the amount of detailed information available, it can be difficult to comprehend the whole system at first. To assist in the development of the initial Data Flow Diagram, several approaches have been formulated. Each approach splits the task of drawing the diagrams into a series of manageable steps. In practice, a combination of the three may be the most appropriate.

Physical document flows

This approach is appropriate if the current system consists principally of flows of information in the form of documents or computer input and output. The initial Data Flow Diagrams are arrived at by:

Listing the major documents and their sources and recipients In addition to actual documents in this list, major information flows of other kinds are listed, e.g. information communicated over the telephone or through computer dialogues. The list compiled for the Yorkies system is:

Booking Sheet
Invoice
Customer Payment
Booking Request
Driver Request and Confirmation
Customer List
Driver Instructions
New Vehicle Documents

Obviously, these are not the full set of information flows within the system, but they are the major ones needed to start the Data Flow Diagrams.

Drawing the document flows   Each source or recipient is represented as an oval in a diagram with the documents represented as flows between them. A source or recipient might be:

- a section within the relevant area of the organization;
- an office in another part of the organization;
- an outside organization or person;
- a computer system.

The relevant sources and recipients for the Yorkies system are initially identified as being:

Local Office Reception Staff
Local Office Booking Staff
Local Depot Staff
Local Office Driver Administration
Head Office
Vehicle Fleet Maintenance
Customers
Drivers
Driver Agencies

These sources and recipients are scattered over a page, and each of the main documents listed before are matched to the relevant sources and recipients. The resulting diagram is shown in Fig. 3.9. In Fig. 3.9 the four different copies of the Booking Sheet are separated to show which copies are used in which areas of the organization

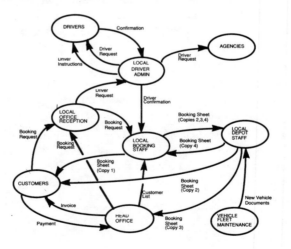

Fig. 3.9

Agreeing the system boundary   The diagram produced is shown to the users for comments on its accuracy and to check that nothing is left out. The boundaries of the investigation may be denoted on this diagram, showing which areas should be included in the investigation, and which should be considered as external to the system. In Fig. 3.9, the sources and recipients that are obviously going to be external to the system are the Customers and Driver Agencies, as they belong to external organisations. Also, the users decide that the activities of the Driver Administration Section should be included, but not the activities of the Drivers themselves, making the Drivers external, but Driver Administration internal. Although an important part of the organization, it is also decided that the Vehicle Fleet Maintenance should be excluded from the system, as this is a large but self-contained section of the organization. The resulting system boundary is drawn onto the diagram as shown in Fig. 3.10.

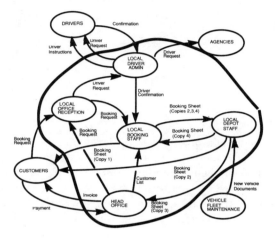

Fig. 3.10

In some projects this definition of system boundary will have been completed in an earlier stage, perhaps as part of an information systems strategy or by a feasibility study.

Identifying processes; within the system   The activities relating to the sending or receiving of the major documents within the system boundary are represented as processes, forming the basis of the top-level Data Flow Diagram. Where these documents are held within files or other means of storage, data stores are added to the diagram. For example, Fig. 3.11 demonstrates how the Booking Request form sent in by the Customers to Local Office Reception and Local Office Booking Staff is now represented as being received by a process, Receive Booking Request, performed by the Reception Staff. The process passes the Booking Request to another process, Fill Bookings, performed by the Booking Staff. The Booking Requests are transformed into Booking Sheets by this process, and these Booking Sheets are put into the Booking Sheets File, represented here as a data store. Note that the section responsible for performing the process is indicated at the top of each process box. This emphasizes the physical mapping of current system functions within the Data Flow Diagrams.

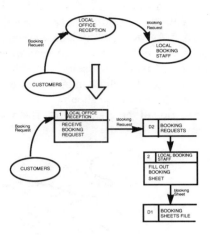

Fig. 3.11

Physical resource flows

This might be an appropriate way of developing the Data Flow Diagrams if the current system consists principally of flows of goods. A Physical Resources Flow Diagram concentrates on following the movement of the physical objects of interest. This flow of goods is represented by broad arrows. The progress of the goods is shown from when they arrive within the boundaries of the system, through the various points at which something is done to them or recorded about them (represented as processes), to their exit from the system.

A diagram that could be used to model the purchasing, rental, and sales of vehicles of the Yorkies system is shown in Fig. 3.12 (this area is actually outside the system boundary in the case study). In this case, the physical resources are vehicles. This shows how the vehicles are tracked from their purchase, through to their eventual sale.

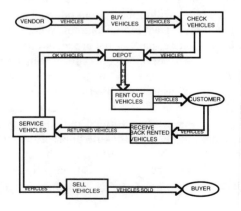

Fig. 3.12

The reason for representing the physical flows is that information will normally flow around the same paths. For example, the vehicle documents will accompany a vehicle when it is bought and sold, and the Booking Sheets will initiate a rental, and be sent when a rental is terminated, as shown in Fig. 3.13. In this way, the information flows and processes of the current system are determined.

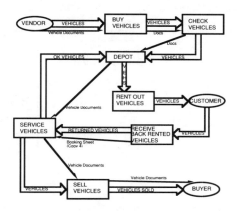

Fig. 3.13

Organization structure

Often, the simplest way to develop a top-level Data Flow Diagram is to use the current organization structure for defining the processes. This approach starts from the point of view of the processes that are performed in the organization, rather than the information that is flowing around the system. When dealing with large systems which have many significant documents and information flows, the physical document flow approach may lead to very complicated diagrams. By looking at the relevant functions of different areas of the organization, a simpler initial picture may be formed, with the detail added later.

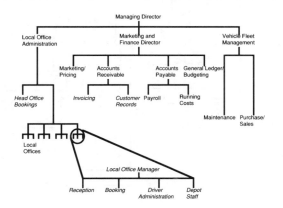

Fig. 3.14

The organization structure of the Yorkies system is shown in Fig. 3.14. Here, the user has decided which of the areas of the organization should be included in the investigation, as shown in italics. The main function of the Local Office Booking staff is to fill the bookings that have been requested by customers. A process Fill Bookings becomes a top-level process on the Data Flow Diagram. By a similar procedure, the main functions of the other areas of the organization are represented as processes, as shown in Fig. 3.15.

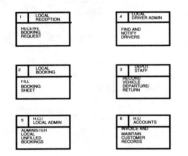

Fig. 3.15

The information flows between these processes, and between the processes and external source/recipients, together with the stores of information are determined and added to the diagram to give the top-level Data Flow Diagram shown in Fig. 3.16. This is the top-level diagram that would have been derived whichever approach had been taken.

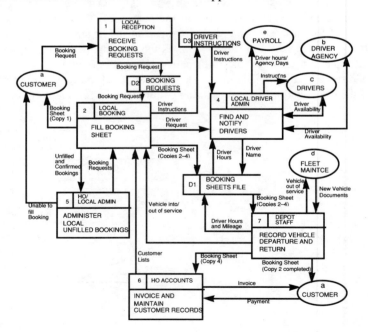

Fig. 3.16

Simplifying complex diagrams

The top-level Data Flow Diagram produced by the approaches above might become very complex and confused. It is important that the top-level Data Flow Diagram is easily understood, as this is a very useful overview of the whole area of investigation, to be used for discussions with users and managers.

As an initial help, if system boundaries and the scope of the system become unclear, it may be useful to draw a context diagram. This is where the system is represented as a single process. All flows into and out of the system are shown around the edge of this process. This is useful in ensuring that the boundaries are correct and well understood. A context diagram for the Yorkies system is shown in Fig. 3.17. Although trivial in this instance, this type of diagram could be very useful for very large projects.

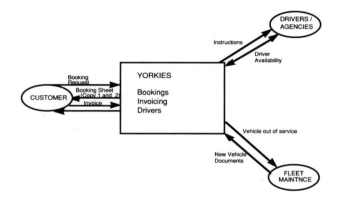

Fig. 3.17

If the top-level Data Flow Diagram is still unclear, then it should be simplified until the overall diagram is intelligible. This simplification may be done in one of three ways (or a combination if appropriate):

1 Combine some of the process boxes, until there are a maximum of 12 boxes on one diagram. The processes that are combined can be split again at a level 2 Data Flow Diagram.

2 Combine data stores at level 1 only, showing them individually at lower levels. This might be done if a number of data stores are holding very similar information: in clerical systems, it is often necessary to hold the same information filed under different headings. The identifiers of the new data stores are 'Dn', where n is a number, and when they are broken down at lower levels, the identifiers of the lower-level data stores are 'Dna', where n is the same number as the top-level data store and a is an alpha character, as demonstrated in Fig. 3.18.

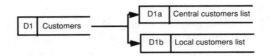

Fig. 3.18

3. Combine external source/recipients at the level 1 only, showing them individually at lower levels. In a system dealing with a large number of different departments of the same overall organization, the name of the organization could be used at the top level, with a breakdown of the departments at a lower level. The composite external source/recipient is denoted in the conventional way at the top level with an alpha character. The individual external source/recipients at lower levels are denoted by the same alpha character and a number. (See Fig. 3.19 for an example.)

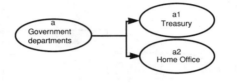

Fig. 3.19

This tidying up of the top-level data flow diagram is shown schematically in Fig. 3.20. The final diagram is less cluttered and much more readable than the initial diagram.

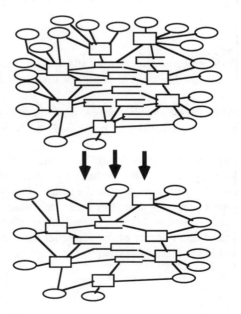

Fig. 3.20

Data flows to and from data stores

Where an update to a data store is required, it is often tempting to show a data flow from the data store to the process in addition to the flow from the process to the data store. It seems to be logical, especially for those from a conventional programming background, that the required information should be retrieved from the data store before it can be updated. However, unless there is some real reason to extract the record first, only the data flow performing the update should be shown. This is demonstrated in Fig. 3.21.

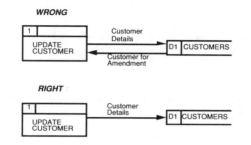

Fig. 3.21

Similarly, if a record is being retrieved from a data store, it is not necessary to show the selection criteria entering the data store before the record being retrieved is output from the data store. This is demonstrated in Fig. 3.22.

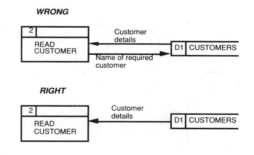

Fig. 3.22

Updates and retrievals

The functions documented on the Data Flow Diagrams are predominantly update functions, where data is being changed in some way. It may appear that this is ignoring an important part of the system in the area of reporting. All but the most major enquiries are omitted from the Data Flow Diagrams to avoid confusing the diagrams needlessly. The others are documented separately on a Retrievals Catalogue. This is simply a list of the enquiries and reports used in the current system. At a later stage, the retrievals documented here may be respecified for the required system or discarded if no longer required.

Decomposition to lower levels

To describe fully the current system in the form of Data Flow Diagrams, it is necessary to expand most of the top-level Data Flow Diagram process boxes to a second level, or possibly to a third or fourth. As a general guideline, processes are decomposed if:

• there are more than eight data flows in to or out of the process;
• the process name is complex or very general, e.g. 'Record Customer Information, Send Invoices, and Receive Payment' or 'Maintain Customer Information'.

At the bottom level, each process should have a brief specific name and have between two and eight data flows surrounding it.

Data flows between lower level processes and other objects should be reflected back up to the top level so that data flows at all levels are shown in summary at the top level. This is demonstrated by the level 2 Data Flow Diagram of process 4 of the Yorkies Data Flow Diagram in Fig. 3.23. It may be seen that all the flows crossing the boundary (shown as a frame) of the lower-level Data Flow Diagram are shown as flows to and from the process box at the higher level.

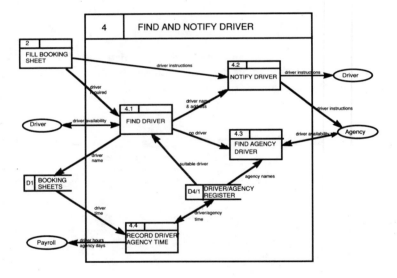

Fig. 3.23

If a data store is used by only one top-level process, then it is said to be internal to that process. To avoid confusing the level 1 Data Flow Diagram, internal data stores are shown inside the frame of the level 2 Data Flow Diagram, and their identifiers reflect the identifier of the top-level process. In Fig. 3.23. the data store D4/1 is internal to process 4 so appears only at the second level. If other internal data stores were shown, they would be identified as D4/2, D4/3, and so on.

If the system is too complex to describe adequately in several levels of the Data Flow Diagram, then Elementary Function Descriptions are used to describe the bottom-level processes. These are brief narrative descriptions of the processes which often cross-refer to existing current system documentation.

SUMMARY
- Current System Data Flow Diagrams are built up by users and analysts jointly.
- Three methods of starting the Data Flow Diagrams are described: Physical Document Flow Diagrams, Physical Resources Flow Diagrams, and using the organization structure.
- A level 1 Data Flow Diagram is built and, if necessary, simplified by combining processes, data stores, and external source/recipients.
- Each level 1 process is expanded to second and, possibly, third-level diagrams, maintaining consistency between the levels.
- Elementary Function Descriptions may be completed for some or all of the bottom-level processes.
- Reports and enquiries are documented on a Retrievals Catalogue.

EXERCISES

CABA—CAt Breeding Agency

CABA (CAt Breeding Agency) specialize in finding mates for pedigree cats. They advertise in magazines such as *Pedigree Cat* and in response to enquiries owners are sent an application form. On the application the owner sends in details of the cat they wish to breed (name, age, sex, pedigree) and their preferences for a mate. Owners may also send in details of their membership of various cat organizations which may entitle them to discounts.

This information is transferred to two card index files—one detailing the cat and its preferred mates, the other holding the owner's details.The original application form is filed by date received.

CABA guarantees to send the owner details of five possible mates. After the cat and owner have been registered the cat card index is examined for potential mates. When the best five mates have been found, details of their owners are extracted from the owners card index and the names of the potential mates are recorded on the cat card. The owner of the cat is then sent details of the five cats and their owners. The owner may make later applications for the same cat if they are unsuccessful—in this case CABA ensure that different mates are found.

Use a Data Flow Diagram to represent this system.

Reckitt Repairs

In order to control the Reckitt Repairs projects there will be a system to track the work and movement of parts. The current system includes a small computer system to track the movement of parts but does not interface this with the control of work.

The types of machine worked on within Reckitt consist of a number of different parts that are removed from the machine, cleaned, and inspected for defects. If defects are found, they need to be remanufactured. Finally all the different parts need to be replaced in the correct machine in the right place. This entire process needs to be monitored and controlled carefully to ensure that parts are not lost and that a project is completed on time and within budget.

Customers send in requests for repairs to the Project Office.

The following data is held manually, with the exception of the parts file:

Job Order Book List of all projects and customers. The status of the project is monitored here under the headings Planned Start Date, Planned End Date, Actual Start Date, and Actual End Date.

Project Plans These are set up by the project office to show the different work packages and the phases within the work packages.

Staff Records A card index of all staff with their qualifications

Work Plans Detailed plans of the work with staff allocations and details of parts to be worked on in a work package.

Machine/Parts Book The machine details and all parts are detailed in this book with any special requirements for any particular part. All movements are recorded here under standard location headings.

Parts File Computer-based file showing the parts and their current and planned locations

Use a Data Flow Diagram to represent this system.

3.4   Investigate system data structure (step 125)

Introduction

This step produces the Logical Data Structure of the current system. It is carried out in parallel with the development of the current system Data Flow Diagrams (step 120) and with the development of the Problems/Requirements List (step 140). This section describes how the current Logical Data Structure is developed with particular reference to the Yorkies case study. Section 2.1 described the components of the Logical Data Structure, how they were related together, and the supporting documentation.

During the first step of SSADM, *initiate analysis* (Section 3.2, step 110), an overview Logical Data Structure was drawn; this was used to agree the area of investigation and to help estimate the resources required for the project. The overview Logical Data Structure would normally act as input to step 125, although in this section the development of the Logical Data Structure from its inception is demonstrated. The end-products of the step are the current system Logical Data Structure supported by Entity Descriptions.

As the step is carried out at the same time as the other information collection steps, the information for the production of the Logical Data Structure should be collected at the same time as that for the Data Flow Diagrams and the Problems/Requirements List. The means of information gathering are the standard systems analysis tasks of interviewing, questionnaires, sampling, document collection, etc. Studying documents used in the current system is particularly useful for identifying data items and entities.

The development of the Logical Data Structure is often best carried out by two or three analysts pooling their ideas together in a 'brainstorming' session. They would then spend some time checking their ideas with users (these users would need to have had some instruction in SSADM) and validating the diagram against the processing requirements detailed on the Data Flow Diagrams. This step should not involve a significant effort, taking about one day on a small project and about five man-days on a medium-sized project.

In this section a systematic approach to developing the Logical Data Structure is described: selecting the entities, identifying the relationships, drawing the diagram, developing the supporting documentation, and then validation of the Logical Data Structure. These are only guidelines to logical data structuring, experienced practitioners will develop an approach that suits them and the particular task best.

Selecting entities

The first step is to identify some of the entities of the current system. Remember that an *entity* is *something of significance to the system about which information will be held and which is capable of being uniquely identified.*

From the description of the Yorkies current system given previously we can identify the following as possible entities:

Customer, Booking, Driver, Invoice, Vehicle, Agency

All of these are things that information is held about in the current system. For instance information about the Customer is held in the Customer List and the Customer File. Note that the Customer File and List are not themselves entities—they are both ways in which the current system implements the Customer entity and both appear as data stores on the current physical Data Flow Diagram. Other current system data stores may have no obvious entity counterpart: the Vehicle Booking Diary and the Empty Vehicle Log use data that spans several other entities.

There are other potential entities that do not match data stores in the current system: Payment, Vehicle Category, Local Office

It is difficult to decide whether these are entities or data items which belong to the entities listed above. Vehicle Category is obviously very important to the system; customers book for a vehicle category, vehicles belong to a vehicle category. But is it an entity itself or does just the Vehicle Category Code belong as a data item in Booking and Vehicle? There are two factors that point to it being an entity:

- There are several data items that could be associated with Vehicle Category: Vehicle Category Code, Vehicle Category Description, Type of Drivers Licence Required, Capacity Range, and others. Vehicle Category Code would uniquely identify each entity occurrence.
- An occurrence of the Vehicle Category entity is related to many occurrences of both the Vehicle and the Booking entities. This is an important point; if there are many occurrences of one candidate entity associated with one occurrence of another candidate entity then both are likely to be entities.

A similar argument could be applied to the Local Office entity.

Payment is a slightly different case. There are several items that we would associate with it, such as Amount Paid, Date Paid, Method of Payment—indicating a Payment entity. However, there seems to be no unique identifier for a Payment and there could be only one payment for each Invoice—indicating that the payment details belong in the Invoice entity.

Resolving this question with the user indicates that a customer may pay for an invoice in several instalments or pay for several invoices with one cheque—suggesting a separate Payment entity.

It is a common mistake to identify a data item as an entity— remember that an entity is made up of data items. Thus Booking No. is not an entity, it is the data item that uniquely identifies an occurrence of the Booking entity.

Another common mistake is to make the organization itself an entity— n this case Yorkies would appear as an entity on the Logical Data Structure diagram with relationships to every other entity. For something to be an entity there must be multiple occurrences of it, and there must be information that would be held for it. In a system like this here is only one occurrence of Yorkies and the information which would be held about it is all the information shown in the Logical Data Structure. The whole Logical Data Structure is Yorkies—it is wrong to show it as a separate entity.

Having identified the entities, we now go on to investigate the relationships between them. Selecting the entities may seem at first like a black art, but with experience it becomes easier. It is not too critical if mistakes are made at this point, new entities are often identified and old ones removed in the later stages of Logical Data Structuring.

Identifying relationships

Each combination of selected entities needs to be examined to establish whether there is a direct relationship between them. Two approaches are described for doing this: a rigorous approach using an entity grid, and an informal approach building the diagram directly.

The entity grid

A rigorous approach which uses a grid to force analysis of every pair of entities and determine the relationship between them. The entities selected previously were:

Customer, Booking, Driver, Invoice, Vehicle, Agency, Payment, Vehicle Category, and Local Office.

These can be drawn up on an initial grid as shown in Fig. 3.24.

Fig. 3.24

The bottom left corner of the grid has been hatched to avoid considering relationships twice; the Customer/Booking relationship would be considered in the Customer row and so has been hatched out to avoid consideration in the Booking row. (There are several possible ways of laying out these grids but each should force analysis of all possible pairs of entities.)

Each box on the grid represents a possible relationship between a pair of entities. So for the top row we consider all the possible relationships that a Customer could have, for the next row we consider all the possible relationships that Booking could have (apart from the one with Customer considered in the Customer row). By proceeding systematically through the grid every possible relationship is analysed.

For each pair of entities the following questions must be asked.

*Are there occurrences of entity A that are **directly** related to occurrences of entity B?* It is important that only direct relationships are identified. For instance, when analysing Customer and Vehicle, we might think 'Customers rent and drive Vehicles therefore there must be a direct relationship between them'. However, the relationship is an indirect one that is maintained through the Booking entity.

If the answer is *Yes* then a further question is asked.

Is that direct relationship of interest to the system? In some cases there may be a direct relationship but one which is of no interest to the system. For example, some occurrences of the Customer entity may have a relationship to other Customer occurrences perhaps as subsidiary companies or as clients. However, this relationship is not of interest to the system (although really only the user can say this).

If the answer to both these questions is *Yes* then a relationship exists between the pair of entities, the appropriate box on the grid is marked with a cross.

Completing the Yorkies entity grid
Below we start by going through the grid systematically although after the first row only examples of particular interest are discussed.

Customer–Customer   This was mentioned above and it was decided that the direct relationship was not of interest.

Customer–Booking   Yes, there is a direct relationship and it is of interest to the system. Customers make bookings and bookings are made by customers. Sometimes a good way of deciding whether a relationship exists and is of interest is whether access is required; in this case we would want to find all the bookings made by a particular customer and which customer a particular booking was for.

Customer–Driver   No, there is no direct relationship—an indirect relationship is maintained through Booking. Note that even if a relationship had been incorrectly identified here further examination of the nature of the relationship or validation of the diagram would have identified and corrected the error.

Customer–Invoice   Yes, customers are sent invoices and this is of definite interest to the system. Sometimes another good way of identifying relationships is to ask if we would hold any data belonging to the other entity in either of the two entities being considered.

Here the Customer No. would be associated with the Invoice, thus indicating the relationship.

Customer–Vehicle   No, considered previously—no direct relationship.

Customer–Agency   No, similar to the Customer–Driver considered previously.

Customer–Payment   Yes, customers make payments which are of interest to the system.

Customer–Vehicle Category   No, although customers book for a vehicle category, this relationship is maintained through the Booking entity.

Customer–Local Office   Customer lists are held at local offices but they include all customers on them. Customers do not deal with just one particular office, they could book at any office. Thus there is no direct relationship only an indirect one via Booking.

Only relationships of particular interest are considered below.

Driver–Payment   Drivers presumably receive a payment for work done. However, the Payment entity in this case refers to a payment made by a customer for a completed booking. It is therefore important to be clear exactly what is meant by each entity name. In this case study the payment of drivers and agencies is not included in the area under study, otherwise a new entity might have had to be introduced.

Driver–Vehicle Category   There seems to be a relationship in that only drivers with certain licences can drive certain vehicle categories. Although this could be an indirect relationship through another entity it does not appear to be via any of the entities identified so far. It is marked in on the grid and considered later.

Local Office–Local Office   There could be a relationship between different Local Offices, for instance one could manage a group of smaller Local Offices. It is marked on the grid with a '**?**' and considered later.

The completed grid is shown Fig. 3.25. The next step is to draw each entity as a box on a diagram and draw in the relationships. Each relationship identified on the grid should be examined to determine its precise nature. This is described in the section, 'Drawing the diagram'.

	BOOKING	DRIVER	INVOICE	VEHICLE	AGENCY	PAYMENT	VEHICLE CATEGORY	LOCAL OFFICE
CUSTOMER	X		X			X		
BOOKING		X	X	X	X			X
DRIVER							X	X
INVOICE					X			
VEHICLE							X	X
AGENCY								
PAYMENT								
VEHICLE CATEGORY								
LOCAL OFFICE								?

Fig. 3.25

Building the diagram directly

This is a more informal approach based upon drawing the entities on a diagram and considering the likely relationships. The disadvantage of this approach is that it does not force consideration of every possible relationship. Its advantage, apart from being slightly quicker, is that it is easier to identify indirect relationships.There is a temptation when building the grid to mark in a relationship between almost every pair of entities—this will lead to a 'spaghetti' style diagram that will be very hard to sort out. However, if only few indirect relationships have been identified these should be easily corrected when a closer examination of the relationships is made and when the diagram is validated.

The relationship between Customer and Vehicle discussed previously was not obviously indirect when the grid was being used. When all the entities are shown on a diagram it is easier to see that the relationship is maintained through the Booking entity, especially if the relationships of Booking with Vehicle and with Customer have already been shown.

The questions asked of the possible relationships are the same whether the grid is used or not:

*Are there occurrences of entity A that are **directly** related to occurrences of entity B?*

Is that direct relationship of interest to the system?

There is really very little difference between the two approaches and personal preference should decide which is used.

In the next section the relationships identified are further examined to determine their nature. When the diagram is built up directly without the use of a grid the identification and examination of relationships are normally performed at the same time. The activities have been separated here because it is easier to explain the rather more formal approach.

Drawing the diagram

After identifying the relationships, the next step is to draw each entity as a box on a diagram and draw in the relationships. There are no precise rules for positioning the entities though it is obviously best to try and place entities close to each other if there is a relationship. The entities identified on the grid are drawn in Fig. 3.26.

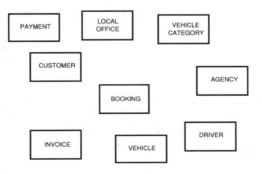

Fig. 3.26

Each relationship identified on the grid should be examined to determine its precise nature. It is necessary to determine the degree and the existence of each relationship (both discussed in Sec. 2.1).

To determine the degree of the relationship between entities A and B we ask two questions:

*Can **one** occurrence of **A** be related to **more than one** occurrence of **B**?*

and

*Can **one** occurrence of **B** be related to **more than one** occurrence of **A**?*

The answers to these questions will determine whether the relationship is one-to-many, many-to-one, many-to-many, or one-to-one.

To determine the existence of the relationship between entities A and B we ask the questions:

*Can an occurrence of **A** exist **without** a related occurrence of **B**?*

and

*Can an occurrence of **B** exist **without** a related occurrence of **A**?*

The answers to these questions will determine whether the relationship is optional.

Taking the first row of the grid we examine the relationships of Customer with Booking, Invoice and Payment.

Customer–Booking

To determine the degree of the relationship we ask:

*Can **one** occurrence of **Customer** be related to **more than one** occurrence of **Booking**?*
Yes

and

*Can **one** occurrence of **Booking** be related to **more than one** occurrence of **Customer**?* No

This means that there is a one-to-many relationship with the many at the Booking end. This is represented by a line with a 'crow's foot' on the Booking entity (Fig. 3.27).

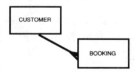

Fig. 3.27

To determine the existence of the relationship we ask:

*Can an occurrence of **Customer** exist **without** a related occurrence of **Booking**?* Yes—
because a customer might not have made any bookings yet but still could be on the Customer List.

*Can an occurrence of **Booking** exist **without** a related occurrence of **Customer**?* No—
because bookings can't be made unless they are for an accredited customer.

This means that the relationship is optional from the Customer (master entity) to the Booking (detail entity) but is mandatory from the Booking to the Customer. As only optional relationships from the detail entity to the master entity are shown in SSADM (this was discussed in more detail in Sec. 2.1) no circle is required on the relationship line between Customer and Booking. Another way to think of this is that in a normal 'one-to-many' relationship the 'many' could be zero.

Customer–Invoice and Customer–Payment
These are both one-to-many relationships with the Customer being at the one end. Neither an invoice nor a payment can exist without a corresponding customer so neither relationship is marked as optional.

Booking–Driver and Booking–Agency
A driver may have many bookings but a particular booking will only be for one driver— this is then a one-to-many relationship with Booking at the many end. A similar relationship exists between Agency and Booking. These two relationships exclude each other in that if a booking is for a driver it cannot be for an agency and vice versa. However, this case is more complex than the simple exclusivity described in Sec. 2.1 in that a booking may not require a driver at all if the customer is supplying his own. Thus a Booking occurrence may be owned by a Driver occurrence or by an Agency occurrence or by neither of them. So both are shown as optional relationships linked by an exclusive arc.

Booking–Invoice
A booking will only appear on one invoice. In this system Yorkies invoice their customers on individual bookings so each invoice will be for one booking. Thus we have a one-to-one relationship between Booking and Invoice. One-to-one relationships are not normally shown in Logical Data Structures since it normally means that the entities can be merged. In this case both Booking and Invoice could be uniquely identified by the Booking No. and the data items associated with Invoice such as Invoice Date and Invoice Amount could be included in the merged Booking/Invoice entity. The two entities are then merged. The relationship Invoice had with Customer is the same as the one that Booking had with Customer and therefore both become simply a Customer–Booking/Invoice relationship.

Vehicle–Booking/Invoice and Vehicle Category–Booking/Invoice
The booking is initially made for a vehicle category and then when the booking is confirmed it is allocated to a particular vehicle. Booking/Invoice is therefore at the many end of both one-to-many relationships. The two relationships are exclusive in that before confirmation the Booking occurrence is related to a Vehicle Category occurrence and after confirmation is related to a Vehicle occurrence. This is shown by an arc linking the relationships.

Booking/Invoice–Payment
The identification of a Payment entity was discussed on pages 83–4, 'Selecting entities'. A customer may pay an invoice in several different instalments. Similarly a payment may cover several invoices. There is then a many-to-many relationship between Booking/Invoice and Payment. This is resolved by the creation of an Invoice/Payment Link entity which is owned by both Booking/Invoice and Payment.

Local Office–Local Office

This was marked with a '?' on the grid for further consideration. If each office was managed by another office then Local Office would have a one-to-many relationship with itself. This would be shown by the crow's leg from Local Office looping back, sometimes known as the 'pig's ear'. In the Yorkies system further discussion with the users reveals that the interrelationships between local offices are not relevant.

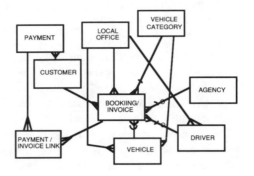

Fig. 3.28

Some other relationships which are 'one-to-many' have not been discussed but are shown in Fig. 3.28. This is Fig. 3.25 with relationships drawn in and the other changes discussed above made. This first draft diagram has many crossing lines and some master entities shown below their details. The layout of Logical Data Structure diagrams was discussed in Sec. 2.1. By repositioning some entities it is possible to avoid all crossing lines and show all masters above their details. The improved diagram is shown Fig. 3.29.

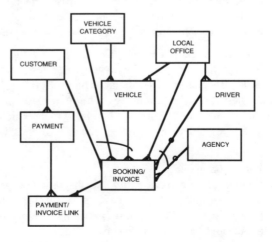

Fig. 3.29

Support ing documentation

The previous section ended with the production of a neat first version of the current system Logical Data Structure diagram. The production of the documentation necessary to support that diagram is now discussed. Each of the components of the Logical Data Structure may be documented: entities on Entity Description forms, data items in the Data Catalogue, and relationships with a Relationship Description. The format of this documentation was described in Sec. 2.1, so in this section we give examples from Yorkies and explain how they were created.

Entity Descriptions
An entity description for the Vehicle entity is given in Fig. 3.30.

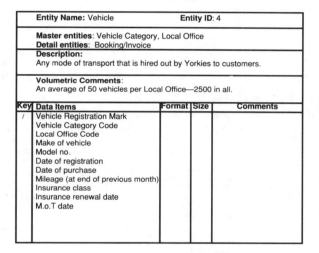

Entity Name: Vehicle			Entity ID: 4	
Master entities: Vehicle Category, Local Office				
Detail entities: Booking/Invoice				
Description:				
Any mode of transport that is hired out by Yorkies to customers.				
Volumetric Comments:				
An average of 50 vehicles per Local Office—2500 in all.				
Key	**Data Items**	**Format**	**Size**	**Comments**
/	Vehicle Registration Mark			
	Vehicle Category Code			
	Local Office Code			
	Make of vehicle			
	Model no.			
	Date of registration			
	Date of purchase			
	Mileage (at end of previous month)			
	Insurance class			
	Insurance renewal date			
	M.o.T date			

Fig. 3.30

Notice that two data items have been included which do not obviously belong with Vehicle: the Local Office Code that the vehicle is registered at and the Vehicle Category Code that it belongs to. These are the notional keys of the Local Office and the Vehicle Category entities. These are appearing as foreign keys in the Vehicle entity indicating the relationships to the master entities. It is advisable to include these foreign key data items in entities as they are required for the data to be normalized (explained in Sec. 6.2) and because users find it easy to understand.

At this stage only the most significant data items would be included with each entity—if any items discovered in analysis obviously belong to a particular entity they should be included; it is silly to discard any information collected. However, an exhaustive exercise to discover all data items and assign them to entities is not necessary.

Data Catalogue
This is a catalogue of all the data items in the system—it may be held on a card index or on a microcomputer. A sample entry from Yorkies is given in Fig. 3.31 for Vehicle Registration Mark.

Data item identifier: 39
Data item name: Vehicle Registration Mark
Data item description: Unique identifier given by the Government to all motor vehicles—used in Yorkies to uniquely identify Vehicles.
Synonyms: VRM, Registration No.
Format: A999 AAA
Range: First character will be in the range A to G
Validation Criteria:
Size: 7 (fixed)

Fig. 3.31

The Data Catalogue would be started in this step and completed as more information is collected during analysis. Documents used in the current system such as completed forms, reports, or files are the main source from which data items are identified.

Relationship Descriptions
SSADM does not require the production of Relationship Descriptions, although they are recommended. A sample description is given in Fig. 3.32 for Booking–Vehicle Category.

Detail Entity Name: Booking	**Entity ID:** 6
Master Entity Name: Vehicle Category	**Entity ID:** 3
Relationship Name:	
Description: Every booking is made for only one vehicle category. This relationship is exclusive with that of Booking with Vehicle, which replaces this relationship when the vehicle is collected.	
Volumetric Comments: Varies greatly depending on the vehicle category. Max 2500, Avg 400, Min 10.	

Fig. 3.32

These Relationship Descriptions would be started in this step using information gathered while analysing the present system.

Validating the Logical Data Structure

When the Logical Data Structure diagram has been produced it should be checked to ensure that no redundant relationships have been shown, that the structure can support the processing required by the current system, and that it is acceptable to the user.

Access paths and navigation of the Logical Data Structure
Relationships indicate access from one entity occurrence to another—it is therefore possible to access an entity occurrence from other entity occurrences even when many other entities and relationships separate them. This is often referred to as *navigation* of the Logical Data Structure and the path taken is often referred to as the *access path*. Consideration of the access paths is an important part of validation of the Logical Data

Structure. It leads to removal of redundant relationships and ensures that the Logical Data Structure supports the processing.

An example of an access path is shown below. Suppose that there was a Yorkies requirement to list the vehicle registration numbers, makes, and models used by a particular customer in the last six months. The part of the Logical Data Structure involved in validating this processing is shown in Fig. 3.33.

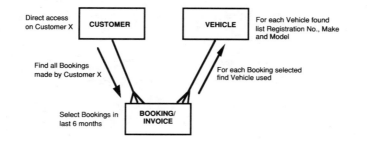

Fig. 3.33

Access path rationalization

The Logical Data Structure should be checked to ensure that no redundant relationships have been created. This is sometimes referred to as access path rationalization since it removes unnecessary or duplicated access paths between entities. Consider Fig. 3.34—there are two paths between entities A and C, one direct and the other through entity B. There are then two ways of satisfying the requirement to find all occurrences of C associated with a particular occurrence of A:

1 Find the particular occurrence of A.
   Find all the occurrences of C using the relationship A–C.

2 Find the particular occurrence of A.
   Find all the occurrences of B using the relationship A–B.
   From the occurrences of B find all of the associated occurrences of C using the relationship B–C.

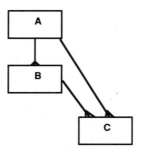

Fig. 3.34

Obviously route 1 is the simpler and quicker route but the Logical Data Structure is not concerned with simplicity or speed of access—it is concerned with ensuring that all the necessary interrelationships are shown. Since route 2 can satisfy the requirement, the relationship A–C is redundant and can be removed. (Note that the relationships involving entity B could not be removed since entity B and its relationships would be required by other processing.)

Inexperienced modellers may introduce many redundant relationships, particularly when a grid is used. Careful analysis of the access paths should remove these.

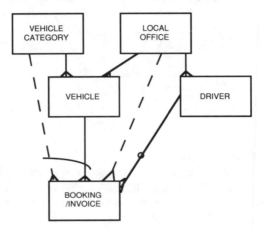

Fig. 3.35

Caution should be used when removing redundant relationships. Consider the part of the Yorkies current system Logical Data Structure shown in Fig. 3.35. The two relationships shown as dashed appear at first to be redundant—if the vehicle category or the local office associated with a particular booking were required this could be found through the Vehicle occurrence associated with the particular booking. This does not consider the changing relationships of the Booking entity over time: a booking is initially made for a category of vehicle at a particular local office—it is only when the Booking Date comes that a particular vehicle is allocated to the booking. Thus the relationship occurrence between a Booking occurrence and a Vehicle occurrence is not created until the vehicle is collected. So if the vehicle category or the local office associated with a particular booking were required before vehicle collection then this enquiry could only be satisfied by using the relationships marked as dashed. Notice that the exclusive arcs shown indicate that the relationships do not both exist at the same time for the same entity occurrence.

Sometimes what appears to be a redundant relationship may not be. Consider the relationship between Local Office and Booking shown in Fig. 3.35. This could be essential if the system allowed a booking to be made at one local office and the vehicle to be collected at a different office. The direct relationship, Booking–Local Office, would indicate the particular office where the booking had been made. The indirect relationship Booking–Vehicle–Local Office would indicate the office from which the vehicle was collected.

Validation against the current system processing
The current system processing will be described by the Current System Data Flow Diagrams and the supporting Elementary Function Descriptions. The current system Logical Data Structure should be checked against the processing to ensure that the access paths can provide the data required for each process.

The validation of the Logical Data Structure is performed each time it is developed or modified. Validation in stage 5 of SSADM is highly formalized with all update or enquiry processing fully documented in terms of the navigation required. In stages 1 and 2 the validation is informal and consists of walking through each bottom-level Data Flow Diagram process and each retrieval against the Logical Data Structure. The informal thought processes that the analyst might have in checking a requirement against the Logical Data Structure are shown in Fig. 3.33.

In stage 1 the processing requirements will be expressed, in their most detailed form, as bottom-level Data Flow Diagrams or as narrative further describing those bottom-level processes (known as Elementary Function Descriptions). Processing that is only enquiry and involves no changes to the data is sometimes only documented as Elementary Function Descriptions. This description of enquiries is often known as the Retrievals Catalogue.

Connection traps
A common mistake is to think of the Logical Data Structure as a railway map—because every station on the railway network is connected to each other it is possible to get from any one station to another without leaving the network. The route may be tortuous but you will get there in the end.

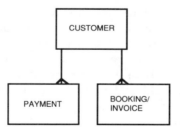

Fig. 3.36

Unfortunately it is not true of Logical Data Structures that navigation is always possible from one entity occurrence to another related entity occurrence. Suppose there is a requirement to match each particular invoice to a payment made by the customer. The Logical Data Structure fragment shown below in Fig. 3.36 appears to satisfy the requirement since Payment is connected to Booking/Invoice through the Customer entity. However, the occurrence diagram (Fig. 3.37) shows that it is impossible given a specific invoice to find the related payment or payments. Given a specific invoice, say Invoice No. 2345, we can find the related customer, Robinsons Ltd. But having found Robinsons Ltd there is no way of determining which of their four payments relates to Invoice 2345. This type of navigation problem is commonly referred to as a *connection trap*.

Fig. 3.37

In this case the way of resolving the connection trap is to create a direct relationship between Booking/Invoice and Payment. As discussed earlier in the chapter this relationship is many-to-many and a Invoice/Payment Link entity is created.

Validation with the user
The Logical Data Structure should be explained to the user. The users should be able to check that all data held in the current system is reflected in the current system Logical Data Structure and supporting entity descriptions. They should also be convinced that the Logical Data Structure can support the current system processing.

Presenting the Logical Data Structure
The users should be involved in the development of the Logical Data Structure. Validation will then take place as the Logical Data Structure develops. However, on most projects a formal presentation of the Logical Data Structure will be necessary at some time. Often on large projects the complete Logical Data Structure is built up from several small Logical Data Structures, each describing part of the system; a formal presentation will then be required to tie the various parts together.

The Logical Data Structure should be presented in such a way as to promote understanding and discussion. Too often users are intimidated by the presentation of a complex diagram showing more than 100 entities with perhaps 200 relationships. The presentation of such a diagram should be carefully planned and the diagram split into smaller chunks for separate discussion. Sometimes the presentation of a summary Logical Data Structure is worthwhile with each high-level 'entity' expanding into a detailed Logical Data Structure.

Validating contents
All data held in the current system should be represented in the current system Logical Data Structure by either an entity or a data item detailed on an Entity Description form. Users should ensure that all data used in the current system is described.

Some projects prefer not to define the contents of each entity until relational data analysis has been performed in stage 4. The Entity Description forms may only detail the major items for each entity. This makes validation of the data content more difficult, since users will need to infer the contents from the entity name. An additional problem is that users often find the concept of an entity harder to grasp than that of a data item. Without engaging in detailed theoretical argument, the authors prefer the earlier assignment of data items to entities—this seems to be general practice in most projects using SSADM.

Users should check that the volumetric information given on supporting documentation is correct.

Validating access paths and relationships

Relationships described on the Logical Data Structure and possibly further defined on relationship descriptions should be checked by the users. The degree and existence of each relationship should be examined to ensure that they represent the user's view of the system data.

The user should be convinced that the Logical Data Structure supports the current system processing. Normally the analyst will 'walk through' some of the more complex or critical navigations as part of the presentation of the Logical Data Structure.

SUMMARY

Step 125, *investigate system data structure,* is carried out in parallel with the development of the current system Data Flow Diagrams (step 120) and with the development of the Problems/Requirements List (step 140).

The major product of step 125 is the Logical Data Structure diagram and its supporting documentation.

Development of the Logical Data Structure involves:

- identification of entities;
- identification of direct relationships between entities;
- creation of a diagram representing the entities and their relationships;
- production of supporting documentation to the diagram;
- validation against the processing requirements;
- validation with the user.

For something to be an entity it should meet the following four criteria.

- It should be of importance to the system being studied.
- There should be information (in the form of data items) associated with the entity.
- There should be more than one occurrence of the entity.
- Each occurrence should be uniquely identifiable

Relationships are identified as follows.

- Relationships should be direct.
- Relationships should be of importance to the system.
- Relationships can be marked on an entity grid or drawn directly onto a diagram.
- The degree of the relationship is determined, if it is many-to-many then a link entity is created or found.
- The existence of each relationship is determined.

The diagram is redrawn with masters above details and avoiding crossing lines.

Supporting documentation to the Logical Data Structure diagram is developed.

- Entity Descriptions which give a definition of the entity, volumetric information, and the data items contained.
- Relationship Descriptions.
- A Data Catalogue is begun in this step; this describes each data item in more detail.

The Logical Data Structure is validated to:

- remove redundant access paths;
- ensure that the structure supports the processing of the current system.

The Logical Data Structure is validated with the user.

EXERCISES

Scapegoat Systems—project management system

1.   An employee can work on several projects at the same time. Each employee belongs to one department and has one manager within the department.
2.   Each project has a start date and a finish date and a number of employees assigned to it. One employee is assigned as project manager. Projects are identified by a Project Code but there is a requirement to list all projects due to finish in a certain week.
3.   Most projects are carried out for a single customer although there are some internal projects for which there is no client. At any one time a client may have several projects under way.
4.   Scapegoat have about 100 customers each identified by a Customer No.

Develop a Logical Data Structure for this system.

CABA—CAt Breeding Agency

Use the description given for the Current System Data Flow Diagram exercise on page 81.

1.   Develop a Logical Data Structure to represent this system.
2.   Describe each entity in terms of a definition and some likely data items (you may need to use your imagination!)
3.   Validate the Logical Data Structure against these requirements:
(a) List all the cats of a particular breed.
(b) Find out which cats (and their owners) a particular cat has been paired with.
(c) Find out how many of CABA's customers belong to a particular club.

Reckitt Repairs

Use the description given for the Current System Data Flow Diagram exercise on pages 81–2.

1.   Develop a Logical Data Structure to represent this system.
2.   Describe each entity in terms of a definition and some likely data items
3.   Validate the Logical Data Structure against the following requirements:
   (a) Find all parts repaired in a particular project.
   (b) Find out which parts are currently at a particular location and which ones will be received in the next day.

3.5   Develop Problems/Requirements List (step 140)

Purpose of the Problems/Requirements List

The Problems/Requirements List is a central document in the analysis stages of SSADM. It is used as a check-list of the factors that must be accounted for in the new system. Each item on the list is a problem or requirement defined by the users. In the later stages of the project this list will be completed by including an agreed solution. The success of a project can be measured by the eventual solutions to each of the problems and requirements on the list.

Sometimes a formal user requirement is produced by the users independently (this often happens as a prelude to involving specialist staff). They are usually expressed purely in narrative and are a useful input into the development of the Problems/Requirements List. However, such user requirements often prejudge the design with such statements as 'I need a terminal linked to a central mainframe', whereas the underlying requirement might be that the user needs data to be up-to-date at all times and will need to be able to access the data during working hours. It is the analyst's responsibility to make sure that the real requirements and problems are identified and expressed logically. It is important to have this 'logical' statement of requirements so that the final solution does not become needlessly constrained. It must be left to the development team to specify the best solution to fit the users' requirements without allowing the users' preconceptions to be carried through to an ill-judged implementation.

Overview of the step

The Problems/Requirements List will have been started in the 'initiate analysis' step. As the current systems investigation continues then further problems and requirements will be identified, usually through interviews with users. The list should be informally reviewed with the users before preceding to the formal quality assurance review at the end of the stage. A part of the Problems/Requirements List for Yorkies is shown in Fig. 3.38.

Problems/Requirements List

System: YORKIES			Sub-system :	Author :		Date :
User name	Prio-rity	Prob ref	Problems/Requirements	Solution		Sol ref
Loc Off Bkg	9	DFD current 2.3	Customers cannot be informed immediately whether vehicles are available for them. If a booking has been referred to Head Office it may take several telephone calls to check on its progress.			
Driv Admin	5	UR 4	Expensive agency drivers are often used in some offices whereas others may not make full use of drivers on their register			

Fig. 3.38

At this stage the analyst is concerned primarily with identifying the problems of the current system. Later in the project, in stage 2, the Problems/Requirement List is further developed by the addition of requirements and eventually solutions. However, if requirements or solutions are suggested by users in the course of interviews during this step they would be included on the form, although they would be considered later. Sometimes the distinction between problems and requirements is rather unclear; the analyst should not be too concerned with this, it is more important to record the item than risk its omission.

Any constraints of the future system should also be identified and added to this list. These may include such items as the environment the hardware will have to operate in, or legal requirements that the system must meet.

The User Name column on the form is used to show the individual or the user area who identified the problem. The Problem Reference column is used to identify any further material that refers to or describes the problem; this might be to a Data Flow Diagram, the Logical Data Structure, a formal user requirement, or to an interview record. The Priority column is for recording the importance of particular items to the users. This can be very useful when various Business System Options are proposed in stage 2; the priorities can be used to evaluate the options. Some projects may use a systematic approach to the determination of priorities similar to the critical success factor approach often used in informations systems planning (see the article by Rockart in Galliers, 1987).

3.6   Review results of investigation (step 150)

The purpose of quality assurance reviews

SSADM lays a great emphasis on holding formal quality assurance reviews at the end of each stage, the procedures for which are described in the introduction to this book. It is very important to ensure that the products from each stage are technically correct and that they meet the objectives of the users. The work for the second stage of SSADM has its foundations in the work done in the first stage. This principle applies throughout the project: each stage builds on the work done in the previous stage. Obviously, with poor foundations, there is a high risk that all subsequent work will be poor.

A formal sign-off by a group, consisting principally of users, emphasizes the joint responsibility for the project of both the users and the project team. This ensures the continuing active interest of the users in the project. It helps avoid the situation often met where communication between the project team and the users is minimal during the development phase leading to the implemented system not meeting the users' requirements.

At the end of stage 1, the most important aspect of the quality assurance review is to gain agreement from the users that the documentation accurately reflects the current system. Thus, it is necessary to ensure the adherence to the basic principles of SSADM without necessarily insisting that the letter of the law is obeyed. The rules of SSADM documentation are more important in later stages.

Stage 1 products to be reviewed

Data Flow Diagrams of the current physical system
These may be supported by Elementary Function Descriptions. Also, if the diagrams are complex, a list of data stores and external source/recipients with a brief description of each might be present. The users should be happy that the areas covered by the Data Flow Diagrams are the correct ones, that none have been left out and that none have been investigated that fall outside the system boundary. The processes shown on the Data Flow Diagrams should be recognizable to the users and the inputs and outputs to these processes should represent real-world flows of information. The technical reviewer might check that flows at all levels are consistent and that there are no anomalies such as data stores being used but not updated.

Logical Data Structure of the current system

This may be supported by brief Entity Descriptions. The users may have more difficulty reviewing the Logical Data Structure than the Data Flow Diagrams, as they may not be able immediately to recognize how the diagram relates to their system. It is important, however, to ensure that the users understand the entities and relationships represented here, as the Logical Data Structure is very central to the analysis and design. The technical reviewer might help by questioning the relationships between entities in the Logical Data Structure.

Entity/Data Store Cross Reference document

Although it is not mandatory to have produced this cross-reference listing at this stage, it provides reviewers with a means of cross-checking the Data Flow Diagrams with the Logical Data Structure: ensuring that all information held within the main data stores is represented on the Logical Data Structure and vice versa.

Problems/Requirements List

The user reviewers will provide the majority of the input here. They will be able to ensure that all the major problems to do with the current system and many of the major requirements and constraints have been considered. It is not important for all of the new requirements to be reflected in this list in stage 1, as the list is not completed until stage 2.

Other documents

Although not necessarily reviewed at the quality assurance review meeting, it is necessary for the project plan and boundary definition to be signed off before the project team proceed to stage 2 of the project.

4. Specification of requirements (stage 2)

4.1 Introduction

Purpose and overview of stage 2

In stage 2, a detailed specification of the required system is built up using several structured techniques and a certain amount of creativity. The specification produced is essentially a logical view of the required system, although it will rarely be entirely free of implementation considerations.

There are significant advantages in specifying the required system in logical terms before deciding how to go about implementing it:

1. It is always best fully to understand a problem before trying to solve it. Consider the design process for a large and complex piece of machinery; before the nuts and bolts are chosen there is a lengthy period where the purpose of the machinery is agreed, plans are drawn up by draughtsmen, and the overall structure of the machine designed. This is so that the final design is not needlessly constrained by the tools used to implement it—by choosing the nuts and bolts before the specification has been properly stated, the best solution may be ruled out in advance.

2. The users can understand a logical specification and are able to verify that their requirements are being met by the new system at an early stage of development—before it is presented to them as a *fait accompli*!

3. If a system is to be developed by external contractors, a logical specification provides all the required information about the system without constraining the developers to particular technical solutions.

4. The earlier errors can be identified, the less costly it is to correct them. The further through the development life cycle a project is, the greater will be the impact of any change in the terms of reference or underlying specification.

5. The maintenance of the system is helped by the presence of the logical specification. The specification can be used as a basis for adding enhancements to the system. If the system is replaced altogether, the logical design can be used as a basis for the design of the replacement system.

Stage 1 of SSADM investigates the current system as perceived and operated by the users of the system. This includes problems experienced by the users and some of the features required of a new system. The Current Physical Data Flow Diagrams are cluttered with the detail of how the current system works as well as what it does. So that the current system is not just reimplemented, some creative thinking is needed. The steps that are involved in going from the current system specification to the logical required system specification are:

- Extract the logical view of the current system.
- Further consolidate the list of problems and requirements.
- Study the constraints that will be imposed on the new system.
- Think about different ways in which the new system might work.
- Build a specification of the required system based upon the above.
- Add to the detail of the specification.
- Check that this is correct and that the user's needs are being met.

Because each step builds on the previous one, several informal reviews are held during the stage in addition to the final review at the end of the stage. This is a very important stage within SSADM. If the requirements are fully sorted out, the remainder of the development will be built on solid foundations. If the requirements continue to change through the remaining stages, the project can be slowed down significantly as the impacts of change increase.

Inputs and outputs of the stage

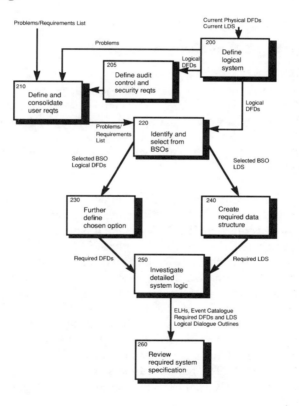

 Fig. 4.1

The products of the current system investigation consist of Current Physical Data Flow Diagrams, a Current System Logical Data Structure and an initial Problems/Requirements List together with any other information gleaned from discussions with the users.

As for stage 2, a major input to this stage is from users of the system. However, the systems analysts will take a more prominent role as they begin to specify the configuration of the new system. The analysts work closely with the users in this stage to achieve a blend of the users' knowledge and experience of the business and the analysts' knowledge and experience in specifying computer systems (see Fig. 4.1 for the structure of the stage).

Outputs from the stage include a consistent set of SSADM documentation describing the required system in some detail. This includes Required System Data Flow Diagrams down to several levels, supported by Elementary Function Descriptions, a Required System Logical Data Structure supported by Entity Descriptions, a completed Problems/Requirements List, Entity Life Histories, an Event Catalogue, and Logical Dialogue Outlines. All of these documents are free of physical implementation factors, although it is unhelpful completely to exclude all references to the eventual configuration of the system.

4.2   Define logical system (step 200)

Purpose of logicalization

In this step, the Current Physical Data Flow Diagrams are re-expressed in the form of a 'logical' view of what the current system does. The Current Physical Data Flow Diagrams are a powerful tool to help analyse the processing and information flows of the current system, but it is difficult to gain a true view of the functionality of the system, as it is embedded in the description of how it is implemented. In order to design a new system, it is important to take a fresh view, looking at what the current system does, independently of how it is achieved.

To be able to look at what the system does, it is necessary to deal with the inessential duplication of function between different sections of the organization. Logically all of the information should be held only once, and it is should be assumed that all of the processes in the system are able to access all of the stored information. In the logical view, the data flows are annotated with only the information necessary for use within the relevant process. Many current systems will have 'evolved', so it is common to find a large amount of duplication and illogical groupings of functions. So, the task of logicalization may require a complete restructuring of the Data Flow Diagrams. Although the guidelines for logicalization are relatively straightforward, this is one of the most difficult steps of SSADM as it requires a complete change of emphasis.

It may be argued that this step belongs in stage 1 of SSADM as it is entirely to do with the current system. The reason it is in stage 2 is that in itself, it adds very little to the current system analysis. Instead, its most useful purpose is to act as a basis for the specification of the required system.

The main purposes of logicalizing the Current Physical Data Flow Diagrams are as follows.

To identify problems within the current system The type of problem identified during logicalization is to do with unnecessary duplication of function and/or information in the current system. Duplication of function means that it is possible that resources are being wasted in doing the same job more than once. Duplication of information, particularly common within clerical systems, means that there is a risk of different versions of the same data becoming inconsistent. Where this information is being used to support decision-making, the impact of this inconsistency may be very great.

To understand the basic functionality of the current system After the completion of stage 1, there will be a very detailed understanding of how the system is currently implemented. This is not always commensurate with understanding the underlying objectives and essential functionality of the system.In producing Logical Data Flow Diagrams, the project team is better equipped to explore ways of supporting the existing environment with a new computer system.

To establish the boundaries of the investigation more closely In stage 1, the areas of the business to be investigated are identified and modelled using Data Flow Diagrams and Logical Data Structures. The logical view of the system allows the identification of the precise functions within each area to be investigated. In practice, this closer definition of boundaries is continued throughout stage 2 of SSADM.

To force a switch in concentration Until now, the investigation has concentrated entirely on the workings of the current system. It is difficult to switch from that viewpoint to a more logical, abstract view. By following the steps in logicalization, the mind is forced away from the 'real-world' view. It is essential to make this mental switch before attempting to specify the required system or it will be difficult to explore the different options available to meet the objectives of the required system: the organization of the old system will persist.

To act as a basis for the specification of the required system Many of the functions of the current system will be required in the new system. The Logical Data Flow Diagrams are used to ensure that the Required Data Flow Diagrams have not missed any essential processes.

Steps in logicalization

Logicalization involves the 'unravelling' of the Current System Data Flow Diagrams to give an ungrouped set of bottom-level processes. Physical aspects are removed from the bottom-level processes and they are rationalized according to specified guidelines. The easiest starting point is to make the data stores logical, relating them to the Logical Data Structure as a way of ensuring consistency between the two diagram types. The processes and data flows are then made logical and grouped to form the top-level processes of the Logical Data Flow Diagram.

Step 1: logicalize data stores

Each data store on the Current Physical Data Flow Diagrams represents either data stored in the permanent base of data (main data store) or transient data that is held for a short time before being used by a process (transient data store) and deleted. Each of the main data stores should be related to the entities on the Current System Logical Data Structure, as this is where the structure of the permanent data is shown. In logicalizing the Data Flow Diagrams, it is desirable to establish a precise relationship between the data stores and the entities on the Logical Data Structure to facilitate cross-validation of the two diagrams. Thus, the rule is that each main data store on the Logical Data Flow Diagram represents a whole number of entities on the Logical Data Structure (one or more) and that each entity on the Logical Data Structure belongs to only one main data store. This rule cannot be applied to the current physical diagrams as information is often duplicated in real life.

The main data stores are derived from the Logical Data Structure. Logically related groups of entities are identified and cross-referenced to a new data store. The best way of documenting this is on a Data Store/Entity Cross Reference form. If there is a problem in identifying the logical groupings of data, it is best to look for entities that are functionally related, i.e. entities that are generally operated on together or that form part of the same major inputs or outputs to the system. Theoretically, it would be possible to create a data store for each entity on the Logical Data Structure, but it is preferable to keep the groupings of data at a more summary level on the Data Flow Diagrams, leaving the close definition of data to the Logical Data Structure. In the Yorkies system the Current System Logical Data Structure is subdivided as shown in Fig. 4.2 to form the main data stores. The resulting Entity/Data Store Cross Reference form is shown in Fig. 4.3.

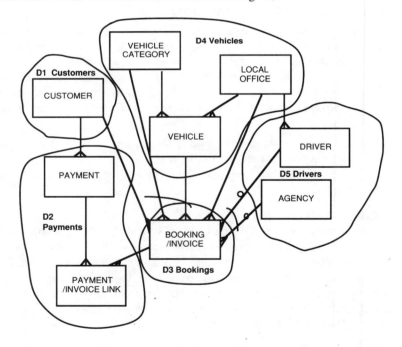

Fig. 4.2

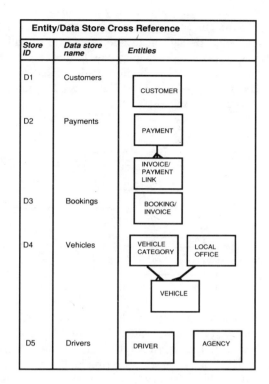

Fig. 4.3

The relationship between the data stores on the Current Physical Data Flow Diagrams and the new 'logical' data stores is shown in Fig. 4.4, which indicates the significant duplication of information on bookings in the current system.

Current system data store		Logical data store	
D1	Booking Sheets File	D3	Bookings
D2	Booking Requests		
D3	Driver Instructions		
D2/2	Vehicle Bookings Diary		
D5/1	Summary Booking Sheets		
D6/2	Invoices File	D3	Bookings
		D2	Payments
D2/2	Local Customer List	D1	Customers
D6/1	Customer File		
D4/1	Driver/Agency Register	D5	Drivers
D7/1	Vehicle History Cards	D4	Vehicles

Fig. 4.4

Transient data stores may exist only because of constraints that exist within the environment of the current system. For example, forms may be batched together by one section before being passed on to another section purely because the internal mailing system only collects the forms twice a day. In this case, there is no logical requirement for a transient data store: logically, the information could be passed directly from one process to another. The Yorkies second-level Data Flow Diagrams show a batching of Confirmed Bookings at Process 2.3 (see Fig. 4.5).

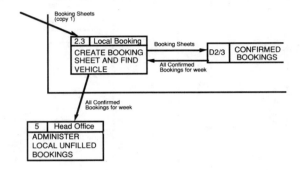

Fig. 4.5

There is no logical reason why this data should be stored before transmission, so this data store is not included in the Logical Data Flow Diagrams. The only reason for allowing a transient data store to remain on the diagram is if the subsequent process needs a complete batch of information all at once. For example, if a process needs to compare a number of records, then it will be necessary to hold them in a transient data store for the whole batch to be read at the same time.

Step 2: logicalize bottom-level processes and data flows

1. A logical process is one that transforms or uses data because the business requires it to do so, independently of how it is implemented. For example, 'Print Report' in the Current Physical Data Flow Diagram would become 'Output Report' and 'Record details in Ledger' would be abbreviated to 'Record details'.

2. Those processes in the current system that are not subject to computerization should be separated from those that will be supported or replaced by a computer. An example of this is where a decision may be made or authorization given only by a responsible person. A process in the Current Physical Data Flow Diagrams named 'Authorize Request' can be treated in one of two ways:

   - If the clerical system surrounding the computer system is within the terms of reference, a clerical process 'Authorize Request' should be separated out from the potentially automatable process 'Record Authorization'.
   - If the boundary of the system under investigation is synonymous with the boundary of the computer system, the person doing the authorizing should be put outside the system boundary as an external source/recipient, with a data flow being shown entering the process 'Record Authorization'.

3. In the Current Physical Data Flow Diagrams, the location where a process is performed is indicated at the top of the process box. All reference to the location is removed in the Logical Data Flow Diagrams. The processes represent what is done independently of where it is being done or who is doing it. The location in a current physical process box may become an external source/recipient. Generally, the external source/recipients shown on the Data Flow Diagrams represent the last to have changed information entering the system or the first to use information extracted from the system.

4. If the data entering a process is unaltered on its exit from the process, the process is replaced by a data flow carrying that information. The process is not transforming the data, merely passing it on or reorganizing it in some way. Also, more minor retrievals are removed from the Data Flow Diagrams and listed on the Retrievals Catalogue.

5. Where several processes are performing exactly the same function, they should be combined. The exception to this is where ambiguity is introduced by doing this or where a large number of additional process-to-process data flows are introduced giving a confused picture. For example, Processes 1 and 2 in the Yorkies Current Physical Data Flow Diagrams both deal with bookings, and may therefore be combined.

6. If two processes are always performed together or serially, they should be combined. For example, one location may perform some initial validation on a piece of information before passing it over to another location for further validation before recording the information on a file. This will have been represented by two processes on the Current Physical Data Flow Diagrams. The Logical Data Flow Diagrams should contain only one process where the two stages of validation form part of the process of recording the information.

7. Where data flows are annotated with document names in the current system Data Flow Diagrams, the actual data items or groups of data items on those forms used by the relevant process should be used to annotate the data flows in the Logical Data Flow Diagrams. For example, in the Yorkies system, only a few items are added to the Booking Sheet initially, so instead of showing the data flow 'Booking Sheet (Copy 1)' as shown in Fig. 4.5, the data items or a group of data items would be shown in the Logical Data Flow Diagrams, e.g. 'Customer Name, Dates Required, Vehicle Type, etc.' However, if certain documents or information flows within the current system are mandated to remain unchanged in the new system, then references to these documents need not be removed from the diagrams.

8. Any processes that must remain clerical should be excluded from the Data Flow Diagrams unless the clerical system is within the terms of reference for the project. In this case, the clerical processes should be clearly distinguished from potentially automatable processes.

9. Often, processes are organized as they are in the current system because copies of the data are held in different places or uncontrolled updating means that checks must be made at regular intervals. Given that all information is now held only once in the logical system, and all access to it is restricted to the processes shown on the Data Flow Diagrams, many of the current system processes may now be unnecessary. For example, in a system where deliveries are made, a copy of a list sent with the driver may be

accepted at the destination and recorded in the files there, but later have to be reconciled with a central register of that information held from the originating depot. This reconciliation should not be necessary logically, as the information is assumed to be available to all processes within the system.

Step 3: group bottom-level processes to form top-level Data Flow Diagram
The top-level Data Flow Diagram should show the logical functional areas of the system. Again, it is difficult to define what is meant here by 'logical'. A useful guideline is that the number of data flows between top-level processes should be minimized, and processes that use the same data should generally be grouped together.

If there is a large number of bottom-level processes to the extent that the logical groupings are not obvious, it may be helpful to draw up a process/data store matrix, showing which processes use which data. In doing this it is possible to identify all of the processes that create, modify, delete, and read the same information. If a group of more than 5–10 processes is identified, then these should be subdivided to give several second-level processes in addition to the single top-level process.

Often in a system, the processes supporting the actual business of the system (such as booking, invoicing, etc.) may be distinguished from those maintaining reference information to support that business (such as maintaining customer information, tables of rules, etc.). These should be clearly distinguished on the Logical Data Flow Diagrams.

The Yorkies top-level Logical Data Flow Diagram is shown in Fig. 4.6.

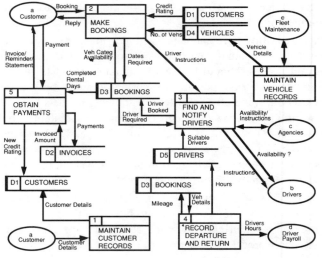

Fig. 4.6

SUMMARY
The steps in creating Logical Data Flow Diagrams that represent the current system functions are:

- Logicalize data stores: main data stores from the Logical Data Structure, some transient ones may be deleted.
- Logicalize bottom-level processes and data flows. All references to how or who removed.
- Group into logical top-level processes. Minimize number of inter-process data flows and group together processes that use the same information.

EXERCISE

Reckitt Repairs

This exercise is based on the Current System Data Flow Diagram and Logical Data Structure provided as answers, in Appendix D, to the exercises in Secs 3.3 and 3.4.

1. Identify logical data stores from the Logical Data Structure, and name them. Produce an Entity/Data Store Cross Reference.
2. Identify probable duplications in the processes on the Data Flow Diagram, and decide which processes are likely to remain inside the boundary and which outside (it should be possible to automate those inside).
3. Draw a logical Data Flow Diagram based upon parts 1 and 2, making sure that the external entities conform to the definition in the text.

4.3   Consolidate user requirements (steps 205 and 210)

Overview

In this section we have combined two steps concerned with the user requirement: step 205, *define audit security and control requirements* and step 210, *define and consolidate user requirements*. Firstly an initial list of audit, security, and control requirements is produced by consultation with the users and other interested parties such as the auditors. This list is then incorporated into the Problems/Requirements List developed in step 140.

The Problems/Requirements List is further modified by including any further problems, requirements, or constraints identified. This list will be used as a checklist for the subsequent specification of requirements.

Factors to be included in the Problems/Requirements List

As the precise configuration of the required system is not decided at this point in the project, the additions to the list will be generalized and independent of the final implementation.

An initial consideration should be given to the audit, control and security aspects of the system. For example, it may be a requirement that certain data is sensitive and may only be accessed by certain users of the system. Similarly, access to certain functions may require a higher level of authorization than others.

The following is a checklist of factors to consider for inclusion in the Problems/Requirements List:

1. BACK-UP AND RECOVERY: Is transaction logging required or is a daily back-up sufficient? How soon must the system be back in service after down-time? How often must a full back-up of the system data files be performed?
2. ACCESS TO THE SYSTEM: Should physical access to the terminals of an on-line system be restricted? Can some users be allowed read access to the data but not write access? Is the system affected by the Data Protection Act? Is there a requirement for several levels of authority or password?
3. ARCHIVING POLICY: How often will data be archived, and how long is it to be retained after archive? Is there a requirement to have access to the data after archive? Is there a requirement to update archived data?
4. ERROR HANDLING: If input is in error, how should this be handled? Should errors in keying of, for example, passwords be notified to a system administrator? Should error data be held for subsequent correction or rejected immediately?
5. AUDIT REQUIREMENTS: Are there any special requirements for establishing audit trails? Are there special reports needed by the auditors?
6. CONTROLS: Are there priorities to establish between different functions or different users? Are check digits required for numerical input or codes?
7. CONSTRAINTS: Is there a physical limit on the size of the computer? Are there specific requirements for cuts in staff or resources? Must the system be live by a certain date? Must the system be compatible with other systems? Must certain software be able to run on the new system?

Quality assurance

The Problems/Requirements List should be expanded and completed without necessarily adding complete entries to the 'Solutions' column. The formulation of the Business System Options, Required System Data Flow Diagrams and Logical Data Structure will help to identify the detailed solutions. As the Problems/Requirements List is such a central document to the subsequent analysis, it is recommended that it is reviewed before continuing to the next step. Part of this review process may include agreeing priorities with and among the the users for the solution of particular problems and requirements. In this way, everyone concerned with the project can be sure that they are following the right and the same track.

4.4   Business System Options (step 220)

Introduction

The Business System Options break away from the current system completely, and begin to explore the shape of the new system. Several possible ways of meeting the requirements of the new system are thought through before the Required System Data Flow Diagrams are started. Very rarely is there only one possible way of designing a system that will meet the requirements of the users. In Business System Options, the creativity of the systems analysts is used to explore the different ways in which a system can be organized, each having its strengths and weaknesses. Obviously, each possible solution will have an impact on the future users of the system, so although the options are devised by the analysts, it must be the users that decide which one to go for.

The Business System Options are not based around physical considerations such as operating system, database management system, etc.—these decisions are made in stage 3 (technical options). Instead, the organization and scope of the system are examined and effects on the working or business of the system are considered. The possible solutions take the form of level 1 (or possibly level 2) Data Flow Diagrams, supported by narrative and Logical Data Structures, and the users select the one that suits them the best in terms of the costs and benefits (both tangible and intangible).

Inputs to Business System Options

The Business System Options are formulated after the Problems/Requirements List has been finalized in step 210. Each of the items on this list is considered in the Business System Options, and if any are excluded, the reasons are documented.

Assuming that the majority of functions present in the current system will be required in the new system, the Logical Data Flow Diagrams will provide an idea of the processing required by the new system.

If the scope of the Logical Data Structure is likely to change as a result of the Business System Options, the Current Logical Data Structure will also be an input to this step.

Any constraints or other factors to be considered should be assembled for the formulation of Business System Options.

The range of the Business System Options

The options presented must obviously differ from one another, but by how much? In some circumstances, very radical changes to the practices and structure of the organization can be suggested, e.g. sack the salesmen and install terminals in the customers' offices. In other circumstances the difference between one option and another can be as small as the extension of the man–machine boundary to allow a small group of users to input their own data.

Some of the factors that determine the breadth of the options offered are:

The terms of reference If the terms of reference are very explicit and limit the project team to, for example, reimplementing the current system without any changes to working practices, the options will be limited. In this case, the user has already specified the outline of the Business System Options before the project was started. Again, the options would be limited where a previous feasibility study has defined the new system in outline. Alternatively, if the terms of reference are broad, the options could consider radical changes to meet the objectives specified for the system in the Problems/Requirements List.

Relationships between users and the project team If the users are keen and involved in the project, they will already have influenced the direction of the required system specification by this time and the project team will have a good idea of what is required. The options will be limited in this case because the decisions are made continuously and gradually.

The nature of the system If the project is providing computer assistance to manual processes that are not going to change, the options will be very much more limited than if the computer is actually taking over some of the processes of the system. In this case, there may be several levels of automation possible (some of which are discussed on page 117, 'Levels of automation').

Development of the Business System Optionss

The suggested approach to developing the Business System Options is to start from the Problems/Requirements List and any other defined constraints that the new system must conform to. There is a definite intention to escape from thinking about the system in the way that it is currently implemented. The analysts think creatively of ways in which a new system could meet the organization's objectives.

The approach to this creative thinking may take the form of a group discussion in which the project team try to invent as many ideas as possible for the new system, including silly ones. This could be followed by a session in which they try to rationalize these ideas into a maximum of six outline Business System Options, expressed as a description and possibly a level 1 Data Flow Diagram. These options would then be informally discussed with the senior users leading to perhaps two or three possible solutions.

In the Yorkies system, there are two major requirements that the new system must meet:

- to improve the operations of the parts of the organization studied in stage 1;
- to be able to deal with one-way hires of vehicles and to track the whereabouts of each vehicle.

There is also a more minor requirement:

- to be able to accept bookings from *ad hoc* (or non-regular) customers.

The Business System Options for Yorkies must be able to meet these requirements. Three such options are described below.

Business System Option 1—centralized system

This involves completely centralizing all the major activities of the company except for the depots for the collection and return of vehicles. All customers would then deal with a Head Office for both bookings and invoicing. Allocation of drivers would also be administered centrally although each driver would only serve a group of nearby depots. Local Offices would be closed down with some staff being transferred to Head Office, some being transferred to the depots, and some being made redundant.

This would involve the purchase of a minicomputer and approximately 35 terminals. Each depot would be informed of their bookings by telephone, and a computer-generated form would be posted to confirm the booking. The mileage covered by the customer would be written on the form, which would then be returned to Head Office. The depots would inform Head Office of any vehicles out of service by telephone.

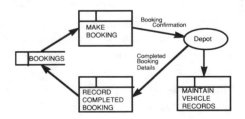

Fig. 4.7

In Fig. 4.7, an extract from the level 1 Data Flow Diagram representing Option 1 is shown. Here, the Depot becomes an external source/recipient, receiving the printed booking form from the system, completing it, and returning the Booking Details to be entered into the Bookings data store by the system. All the processes shown here would be performed on the central computer at Head Office.

Option 2—local autonomy for the offices

This is almost the complete opposite of Option 1. Each Local Office is responsible for its own bookings, drivers, and invoicing. Information is held centrally and shared by all offices Almost all data is entered at the Local Offices. Thus customers would deal with their local office for both bookings and invoices. Each office can make a booking at another nearby office if it cannot be satisfied at the originating office.

This option might require the purchase of a minicomputer capable of supporting up to 60 terminals, communications hardware and software, about 55 terminals or microcomputers (one for each office and some for Head office), and about 52 printers.

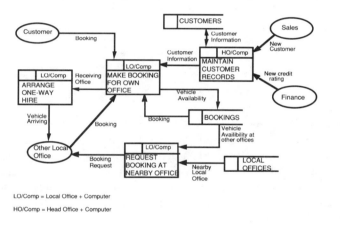

LO/Comp = Local Office + Computer

HO/Comp = Head Office + Computer

Fig.4.8

Figure 4.8 shows the parts of the system where Option 2 differs from the Logical Data Flow Diagram (Fig. 4.6). This shows the distributed nature of the system, with Head Office responsible for the Customer Records and the Local Offices responsible for the bookings. Another way of showing the distribution would have been to draw separate Data Flow Diagrams for Head Office and Local Offices.

Option 3—a distributed system

This is intermediate between Options 1 and 2. In this case Yorkies would be split into five regions each controlled by a Regional Office. Head Office would also become a Regional Office and the four largest Local Offices in each region would become Regional Offices. In each region five of the Local Offices would be closed down, leaving four Local Offices and ten Depots for each Regional Office to administer.

The Regional Offices would hold all the information for the region and be responsible for invoicing. The Local Offices would enter the bookings and add the mileage and driver time when completed. As in Option 2, each office can make a booking at another nearby office if it cannot be satisfied at the originating office. Communications facilities with other Regional Offices would allow one-way hires to be handled.

Each Regional Office would have a minicomputer and two or three terminals or microcomputers. Each Local Office would have one terminal or microcomputer linked to the Regional Office machine and a printer.

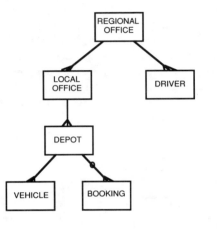

Fig. 4.9

As before, a Data Flow Diagram would demonstrate how Option 3 would work. This would be very similar to the Logical Data Flow Diagrams. Modifications to the Logical Data Structure result from this option. These are shown in Fig. 4.9, where it can be seen that a Regional Office is responsible for several Local Offices and that each Local Office is responsible for several Depots. Vehicles and Bookings are associated with a Depot.

A large number of options could have been developed for this system, each varying slightly in its central, regional, and local flavour. The options presented here demonstrate the two extremes and a middle ground—the user need not choose any of these options. Instead a combination of mainly one option and aspects of the other options could be selected.

Differences between the Business System Options

The range of Business System Options was discussed on page 113. This section describes the ways in which different options can differ from both the original system and the other options. These differences can be categorized as:

- distribution of the system;
- system boundaries;
- levels of automation.

Distribution of the system

In the Yorkies system, the options have been mainly concerned with the distribution of the system. If chosen, both Options 1 and 3 would have meant a significant restructuring of the company. Distributing or centralizing systems will almost always have a major impact on an organization, the effect of which will need to be carefully considered by senior management.

System boundaries and man–machine boundaries

The boundaries of the investigation are originally set in stage 1, now in this step in stage 2 we set the boundaries for design. Each option should have both system and the man–machine boundary clearly defined (this does not mean that they may not be changed later).

If the Yorkies microcomputer which produces the invoices and customer lists in the current system was considered to be satisfactory, then one possible option would have been to leave the invoicing and customer records outside the boundary of the investigation. In this case, an interface with the new system would have to be specified. As this could be a possibility with both Options 1 and 2 it could be regarded as a sub-option of both.

Within the the system boundary, the man–machine boundary may be defined. Two examples of possible man–machine boundaries are described here:

- The outgoing and incoming mileages are recorded within the Yorkies system when a vehicle is rented out and returned. The mileage could be entered directly into the system by the depot staff or they could fill in a form for input at the Local or Head Office.
- In a stock control system, it might be suggested that hand-held devices are carried by the person performing a stock count when determining actual stock levels. This would necessitate the use of computer equipment by relatively unskilled workers. Factors to be considered here would be impacts on wages, trades unions, retraining and other related matters.

All of these implications must be made clear so that the user can decide upon the exact boundaries of the computer system.

Levels of automation

In developing computer support for any process, several levels of automation can be defined. For example, a stock control system would require some way of ensuring that stock levels remain above a certain level. A computer system could deal with this in three basic ways each offering an additional level of automation.

- The most basic option would be for a report to be produced every week showing all the current stock levels, the warehouse clerk would then raise an order for the low stock items.
- A prompting option would determine daily whether any stock levels had fallen below the prescribed level and would generate a report to the warehouse clerk telling him to reorder.
- A fully automatic option would determine when stocks had fallen below the prescribed level and automatically reorder from the suppliers offering the best terms.

Presentation of the Business System Options

The form that the presentation of the two or three favoured options should take will depend on the project. If the project is large, with several groups of users, then it may be necessary to produce a report and have a formal presentation. In small projects or those where the users are closely involved, informal discussions could be sufficient.

In order to make an informed decision the following information should be available to the users:

A description of each Business System Option

This should comprise a level 1 (possibly supplemented by a level 2) Data Flow Diagram, supported by narrative descriptions of the system's functions. If the system's Logical Data Structure is changed by the option then this should also be available. Similarly if there are organizational implications of the option then they should be described.

An idea of the costs and benefits of each Business System Option

It will not be possible at this point to do a detailed cost/benefit analysis. Some estimates and comparisons can, however, be made on which a decision can be based.

As part of the presentation a table could be drawn up listing the relative advantages and disadvantages of each option. These could be referred to items in the Problems/Requirements List.

Selection of a Business System Option

The users are invited to select one or a combination of the features from several of the options presented. This selection then becomes the basis for the full specification of the system in the subsequent steps.

As a result of the discussions at Yorkies the users decided to pick Option 2 although elements of the other two options were selected for combination into the required system specification. The chosen combination is described below.

- The organization of the company remains the same as the current system with no offices being closed.
- Information is to be stored centrally on a minicomputer
- Each Local Office will have a terminal or microcomputer (to be decided at stage 3) and printing facilities.
- Head Office will remain responsible for customer records and invoicing.
- No terminals will be installed in the Depots. As the Depots adjoin the Local Offices, printed booking forms will be used to transmit information to and from the Depots.
- Each Local Office will accept bookings for any office.
- Local Offices will be able to modify any non-financial information about the customer and will validate bookings against the customer records.
- Drivers will be organized into regional pools so that each driver might be used by any one of the offices in the region

SUMMARY

Business System Options:

- take a fresh creative view of the required system;
- are based on the Problems/Requirements List;
- the 'brainstorming' approach is often used to generate many possibilities;
- the possibilities are narrowed down to two or three options for presentation to the user.

Each option is:

- expressed as level 1 Data Flow Diagrams;
- possibly supported by Logical Data Structure, narrative, and cost/benefit analysis.

Options can vary in:

- distributed nature of new system;
- levels of autonomy;
- position of system boundary.

Users pick one (or a combination) option

4.5 Further define chosen option (step 230)

Objective

The objective is further to define the chosen Business System Option. The processing of the required system is is specified by:

- a set of Data Flow Diagrams ;
- descriptions of the major update and retrieval functions;
- descriptions of the content of the data flows across the system boundary.

Approach

This step is done concurrently with the development of the required system's Logical Data Structure (step 240). The two steps are closely linked in that the Logical Data Structure must be able to support the required system processing and the main data stores must correspond to entities on the Logical Data Structure.

The level 1 Data Flow Diagram, representing the chosen Business System Option, forms the basis for the definition of the required systems data flows. The level 1 diagram is expanded to further levels using the Logical Data Flow Diagrams and ensuring that solutions to all entries in the Problems/Requirements List are included. If necessary the processes may be further described using narrative. The Data Flow Diagrams are cross-checked against the required Logical Data Structure by use of an Entity/Data Store Cross Reference form. Processes are also cross-checked by ensuring that the necessary access paths are on the Logical Data Structure.

Descriptions of all the update functions, grouped into batch and on-line, form a document called the Function Catalogue. Update functions can be identified from the required Data Flow Diagrams and will each form a grouping of several low-level processes. Descriptions of all enquiry functions form a document called the Retrievals Catalogue. This is initiated during the current system investigation and is refined during the analysis phase by referring to entries related to information retrieval in the Problems/Requirements List and by discussions with users.

The inputs and outputs (screens, prints, etc.) of the system are described in terms of their constituent data items. These are referred to as I/O Descriptions and may be accompanied by draft formats (e.g. screen layouts, report formats). Thus I/O Descriptions will be developed for every data flow crossing the system boundary and for reports and enquiries detailed in the Retrievals Catalogue.

These three major products of the stage—the Required Data Flow Diagrams, the Function Catalogue, and the I/O Descriptions—are described in further detail using examples from the case study in the next sections.

Required System Data Flow Diagrams

In the previous step Business System Options were developed and represented as level 1 Data Flow Diagrams. One, or a combination, of these was chosen by the users for development into the new system. The first part of this development involves the production of a full set of Data Flow Diagrams.

In the Yorkies case study a combination of mainly Option 2 (local autonomy) with some elements of Option 1 (centralized system) was selected. The relevant elements are taken from the appropriate Business System Option diagrams to form the level 1 Required System Data Flow Diagram. This is shown in Fig. 4.10. Process 3 of Fig. 4.10 interacts with the Depot as an external source/recipient as in the Option 1 Data Flow Diagram (Fig. 4.7). The chosen solution has a terminal in each Local Office (from Option 2) which are used to make the bookings.

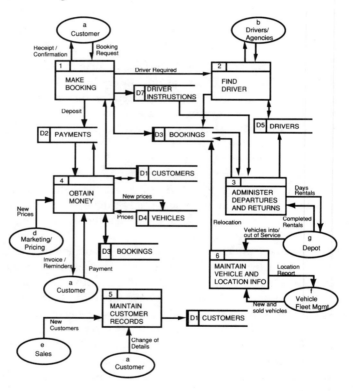

Fig. 4.10

The Data Flow Diagrams should include appropriate solutions for entries on the Problems/Requirements List. This may mean that some additions have to be made to the diagrams. For example, the requirement for accepting unknown customers involves obtaining a deposit from them, issuing a receipt, and recording the payment. This is shown on both the level 1 diagram (Fig. 4.10—Process 1) and on the level 2 diagram (Fig. 4.11— Process 1.1).

Decomposition of the Required Data Flow Diagrams follows the same principles as in the current and logical diagrams, i.e. processes that are complex, cover a multiplicity of functions, and deal with a large number of data flows should be decomposed.

The development of the lower-level Data Flow Diagrams should follow the spirit of the chosen Business System Option. However, in some areas the chosen option may not differ very significantly from the logical view. In this case, processes from the logical diagrams can be incorporated directly. The logical processes Obtain Money, Maintain Customer Records, and Maintain Vehicle Records are substantially unchanged by the Business System Options step and are shown in a very similar way in the level 1 and level 2 Required Data Flow Diagrams. It is worthwhile going through the Logical Data Flow Diagrams systematically to ensure that all processing required from the old system is included in the new.

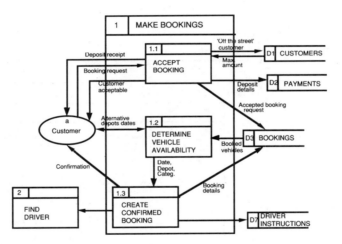

Fig. 4.11

The definition of detailed manual procedures is dependent on the physical implementation and therefore purely manual procedures are rarely shown on Required System Data Flow Diagrams. However, in the Yorkies system a printed list of the day's departures is manually combined with the driver instructions. This is shown in Fig 4.12.

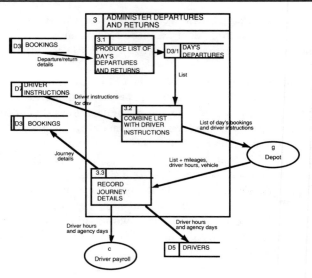

Fig. 4.12

Elementary Function Description		
System : YORKIES	Sub-system :	Author : Date :
Proc ID	Process Name	Description
1.1	Accept Booking	Booking requests are received by telephone or by post. If by telephone then 1.1, 1.2, and 1.3 should be completed in the duration of one telephone call. Driver instructions are noted down from the telephone conversation or received with the booking request. If the customer is known to be an accredited customer then the system checks the maximum value they can book for. If the booking is below that level it is accepted. If above they are asked to provide a deposit or to refer the matter to Head Office customer enquiries. Non-regular customers, those who are not registered on the customer list, may apply to Head Office customer enquiries to become an accredited customer. Bookings will always be accepted from non-regular customers if they provide a deposit of more than the estimated cost of the booking. If a deposit is received this is recorded, a receipt issued and the booking accepted.
1.2	Determine Vehicle Availabilty	The Bookings file is checked to see if vehicles of the requested category are available at the preferred depot on the dates required. If none is available then the nearest depot is found that can fill the booking. The customer is contacted by telephone to check whether this is acceptable. Alternative dates may be sought until a satisfactory alternative is found.
1.3	Create Booking	When a suitable booking has been found the Booking record is created. A confirmation is printed and sent to the customer. The Booking No. is written on the driver instructions. These are then filed manually by date of vehicle departure. If the booking is to start at another depot then the instructions are posted to that depot.

Fig. 4.13

Elementary Function Descriptions are descriptions of the bottom-level processes. The Elementary Function Description can be expressed as narrative, decision tables, mathematical formulae, structured English or as a combination of these. However, it is important to note that SSADM does not, unlike some other methods, define the detailed processing by decomposition of the Data Flow Diagrams. The temptation to get too detailed in these descriptions should therefore be avoided. As an example the Elementary Function Descriptions for Processes 1.1, 1.2, and 1.3 are shown in Fig 4.13.

The Required Data Flow Diagrams are informally checked against the Required Logical Data Structure to ensure that the data and access paths necessary for each process exist. This is documented as an Entity/Data Store Cross Reference (as in the Logical Data Flow Diagrams step). This checking is described in Sec. 4.6.

Other supporting documentation to the required Data Flow Diagrams that might be produced are:

- an External Source/Recipient List and Descriptions;
- a Data Store List and Descriptions (this is additional to the Entity/Data Store Cross Reference which should always be produced);
- a Data Flow Inventory listing for each data flow: the source and destination, the names given on the diagram, and possibly the data items that make up the data flow.

Problems/Requirements List			
System : YORKIES	Sub-system :	Author :	Date :
Prob ref	**Problems/Requirements**	**Solution**	**Sol ref**
DFD current 2.3	Customers cannot be informed immediately whether vehicles are available for them. If a booking has been referred to Head Office it may take several telephone calls to check on its progress.	A local office can book at other offices and immediately confirm over the telephone where a booking has been made.	DFD required 1.3
REQ 1	The system should be able to deal with 'unknown' customers.	Accept 'unknown' customers if they pay a deposit greater than the estimated cost of the rental. Any difference will be resolved when the customer is invoiced.	DFD required 1.1

Fig. 4.14

As was said before, the entries on the Problems/Requirements List should be solved during this step and during the Required Logical Data Structure step. While developing the Required Data Flow Diagrams an eye should be kept on the Problems/Requirements List to ensure that every problem and requirement that is related to updating the system data is considered and a solution found. The Problems/Requirements List should be amended to include the appropriate solution. Figure 4.14 shows solutions added to the Problems/Requirements List after completion of the Required System Data Flow Diagrams.

The Function Catalogue

The Function Catalogue is divided into three parts: on-line update Function Catalogue; batch update Function Catalogue; Retrievals Catalogue

These catalogues have several uses:

- They explain at a high level using simple English what the system can do. thereby making it easier for the users to understand what the new system will be like.
- They form the groupings of low-level processes and retrievals which will be handled by each program of the new system.
- They show whether functions will be dealt with in batch mode or on-line.
- They are the primary means of specifying the retrievals of the new system

Function Catalogue

System : YORKIES	Sub-system :	Author :	Date :	(B)O/R

Funct No.	Function name	Assumes proc Nos.	Description
1	Produce Invoices	4.2	Every Friday the completed bookings for the week are invoiced. Any deposits paid by customers are deducted from the invoiced amounts—in some cases this means a repayment is due.
2	Produce Reminders	4.4	Every Friday reminders are printed for all the invoices sent more than 4 weeks previously which are unpaid. Further reminders are sent (up to 3) for each 4-week period before legal action is taken. Reminder dates are recorded on the invoice record.

Function Catalogue

System : YORKIES	Sub-system :	Author :	Date :	B(O)R

Funct No.	Function name	Assumes proc Nos.	Description
4	Make Booking	1.1, 1.3	This is carried out daily on demand as customers send in or make telephone bookings. Each period for which a vehicle is required is assigned a vehicle of the required category, a depot from which it may be collected, and the dates for the Booking.
5	Find Driver	2	After function 4 has been carried out then if a Booking requires a driver, one is found using function 5, this is also carried out daily on demand. It involves first trying to find a full-time driver to fulfil the rental, then if none can be found at that depot trying to find freelance drivers (which will require telephone calls) and in the last resort using the services of an agency.
6	Record Journey Details	3.3	This is carried out at the end of each day when the depot returns the completed list of the days rentals. Departure and return mileages, and driver time are entered into the system.

Fig. 4.15

The Update Function Catalogue

Each update function is identified from the Required System Data Flow Diagrams. Starting from the level 1 diagram, each process is examined in turn. If a process is entirely batch or on-line and contains only update or retrieval processes, it is directly mapped to a function on the appropriate Function Catalogue. If the level 1 process is a mixture of batch and on-line, update and retrieval, the second-level diagrams are examined and groupings of processes identified as functions and added to the appropriate catalogue. There are no precise rules about doing this, but some guidelines are given.

1. For batch functions:

Processes grouped together must be dealt with in the same time frame (e.g. daily, every Friday, end of month, end of financial year, end of calender year).

2. For batch and on-line functions:

Processes grouped together should update the same entities.

Processes that carry out the same function on different entities (e.g. housekeeping functions) should be grouped together.

3. For on-line functions:

Processes should be grouped together by the people who will use them (e.g. Accounts, Depot, Local Office) because an on-line function may well form a high level menu option.

Some examples from the case study are given in Fig 4.15.

Function Catalogue					
System : YORKIES	Sub-system :		Author :	Date :	B/O/R
Funct No.	Function name	Assumes proc Nos.	Description		
3	Departure/ Return List		This is printed out every morning at each Local Office. It lists all the departures and returns that the depot should deal with that day. For each departure the Vehicle Category is given, the particular vehicle used is selected by the depot staff. The processing for this retrieval will need to be carried out as part of an overnight batch run.		
6	Depot Vehicle Usage Report		This is produced at Head Office every month. It shows for every local office, for each Vehicle Category, the number of days vehicles have been available of the category, the number of days booked, and the number of days out of service.		
7	Vehicle Location Report		This is printed every week at Head Office. It shows for every depot the number of vehicles in each category located at that depot at the end of that week.		
12	Booking Enquiry		Customers may phone up at any time to enquire about a booking; such as details of the driver, what the address is of the departure depot, etc.		
13	Vehicle Availability Enquiry		This is required on demand as part of the Make Booking process. The availability of vehicles of a particular category at a specific depot on certain dates will have to be found.		

Fig. 4.16

Retrieval Catalogues

A retrieval function extracts information from the system's data without making any changes to it. This information may be presented on a screen as a response to an *ad hoc* enquiry or may be printed as a report out of the batch processing cycle. A list of retrievals will have been started when describing the current system processing and further modified during the development of Logical Data Flow Diagrams. Retrievals may also be the subject of entries in the Problems/Requirements List. They are now, like the update functions, divided into an On-line Retrievals Catalogue and a Batch Retrievals Catalogue. These catalogues do not group together similar activities as the update Function Catalogues do;

each entry is for a separate enquiry. Figure 4.16 shows parts of the two Retrievals Catalogues for the case study.

Every retrieval should be checked against the Logical Data Structure to ensure that the data and access paths will exist to satisfy the retrieval. This checking is discussed in Sec. 4.6, which deals with the Required Logical Data Structure.

Input/Output Descriptions

A description of the data content of every input to and output from the system should be developed. This is called an Input/Output Description and is usually shortened to I/O Description, which is the term used in this book.

The I/O Descriptions will be refined as the development proceeds and will form the basis for the detailed design of screen, reports, and forms necessary for program specification. In SSADM these descriptions are also used as input to relational data analysis in stage 4, which is used to define the files or database structures of the new system.

The users of the new system must be deeply involved in the development of I/O Descriptions. They must ensure that the data content is correct and consistent with how they see the new system operating. Some help in defining the precise content of input and outputs of the new system may come from those of the old system.

The inputs and outputs concerned with updates can be determined from the data flows on the bottom-level Data Flow Diagrams. Some projects, particularly those using computer-based tools, will have recorded the data items associated with each bottom-level data flow. These data flow contents will then form the basis for the I/O Descriptions. Otherwise it will be necessary to develop the I/O data content from the information on the data flow lines and from discussions with the users.

The inputs and outputs concerned with retrievals will have to be developed from the Retrievals Catalogue and with the help of the users.

Sample formats for the inputs and outputs may also be started in this step, although the full development of these is done when the nature of the hardware and software is known in stage 6 (physical design). These could be sample forms, report layouts, or screen formats. The use of prototyping tools for developing screen layouts and man–machine dialogues is discussed in Sec. 4.7. Starting development of the man–machine interface at this point has the advantage that it involves and encourages users but it may be wasted effort if the eventual hardware and software differ greatly in their man–machine interface from that of the prototype (machine or paper based).

An example from the case study is shown in Fig. 4.17.

I/O Description				
System : Yorkies	Sub-system :		Author :	Date :
Name: Booking Request		Ref No.: a-1	Type: Form / screen	
Description: Customers send in a Booking Request Form or may give the same information over the telephone. In either case the information is entered into the system on-line to create the booking.				

Data items	Format	Length	Comments
Customer No. Customer name Vehicle Category Code Required Booking start date Required Booking finish date Driver requirement (Y/N) Office No. (Start) Office name (Start) Office No. (Finish) Office name (Finish)			The next group of data items may be repeated if there are several Bookings { only inserted if { one-way hire

Fig. 4.17

In the above example the I/O Ref No. given is a data flow identifier showing that the source of the data flow is external source/recipient 'a' and the destination is process 1.1. If it had been for a report then an identifier related to the retrieval number would be appropriate. 'Type' refers to the physical form the input/output takes; this example is an input form or a screen. The data items listed should be consistent with those included on the Entity Descriptions; maintaining consistency can be difficult when manual documentation methods are used. 'Format' and 'Length' refer to the data items and can be included if defined previously. 'Comments' are related to the inclusion of the particular data item on the input or output.

SUMMARY

→ The required system is specified in terms of its inputs, outputs, and processing.
→ This step is carried out in parallel with and cross-checked against the development of the Required System Data Structure.
→ The chosen Business System Option is expanded into a full set of Required Data Flow Diagrams and supporting documentation.
- Lowest level processes are grouped into Batch and On-line Function Catalogues.
- Enquiries and retrievals that do not change the system's data are described in Batch and On-line Retrievals Catalogues.
- The Problems/Requirements List is amended to include solutions devised in this step.
- For every input and output from the required system an I/O Description is developed defining its data content.

4.6   Create required system data structure (step 240)

Introduction

In this step the Required System Logical Data Structure and its supporting documentation are developed. This involves extending the Current System Logical Data Structure, previously developed in step 125, to incorporate the new requirements detailed in the selected Business System Option (see Fig. 4.18).

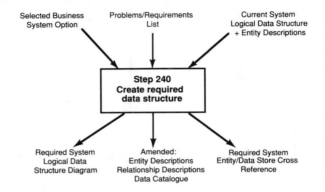

Fig. 4.18

The step is carried out concurrently with the definition of the required system processing (step 230—*further define chosen option*). An Entity/Data Store Cross Reference is developed—each required system data store will contain one or more required system entities. All processing defined is checked against the Logical Data Structure to ensure that the necessary navigation is supported.

The outputs of the step are used extensively in the forthcoming steps of SSADM. An Entity Life History is developed for each required system entity in step 250, *investigate detailed system logic*. The Required System Logical Data Structure is compared with the results of relational data analysis in stage 4 to produce the Composite Logical Data Design.

Modelling the new requirements

The Current System Logical Data Structure developed in stage 1 supports the processing performed in the current system. To satisfy new requirements, changes to the current system data structure may be necessary. These new requirements will come from two sources: the selected Business System Option and the Problems/Requirements List. Both should be examined to identify any necessary changes to the data structure. These changes may involve addition of new entities and relationships to the Current System Logical Data Structure diagram or only minor changes to the supporting documentation. The development of the Yorkies Current System Logical Data Structure into the required view is described below.

New structures

One Business System Option requirement is that the freelance drivers be organized into regional groups thereby increasing flexibility in driver allocation and reducing the use of expensive agency drivers. This results in a new operational master: Driver Region. It is master to Local Office because a driver region will service a group of local offices. The relationship with Driver is exclusive with the relationship of Local Office with Driver because a permanent driver will be attached to an office whereas a freelance driver will belong to a region.

The Problems/Requirements List detailed that the new system should be capable of administering one-way hires and the tracing of vehicles. This leads to some additional relationships on the Logical Data Structure.

A one-way hire indicates that the starting local office will be different from the finishing local office. There are then two relationships between Booking and Local Office: one indicating the *from* office and the other indicating the *to* office. Another change is in the nature of the relationship of Vehicle with Local Office; in the current system this indicated the base location of a vehicle and in the required system it indicates the office at which the documents for a vehicle are kept (see Fig. 4.19).

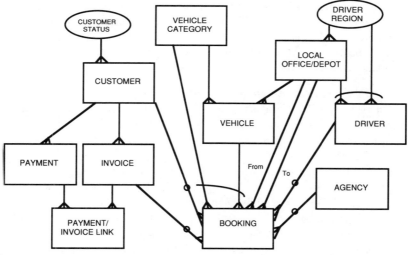

Fig. 4.19

The new system's acceptance of one-way hires causes another problem; some destinations would prove more popular than others so vehicles could get stuck in undesirable locations. To solve this problem a new Process 6 'Maintain Vehicle and Location Information' was shown on the required Data Flow Diagram (Fig. 4.10). One of the activities within this process is the relocation of vehicles which have become 'stuck' in undesirable locations. This relocation is managed within the new system by means of an internal one-way booking. The internal booking will be very similar to the ordinary booking except that there will be no corresponding customer and no invoice. The internal and external bookings both belong to the same entity type Booking. The internal booking makes the relationship of Booking with Customer optional.

In the current system one invoice is created for each booking and so the Invoice and Booking entities were merged. A requirement for the new system is that each customer is invoiced (provided they have had bookings) every two weeks. This means that an invoice can be for many bookings and therefore to the separation of Invoice and Booking as shown. The relationship is optional because a booking will be created before the invoice and must then be able to exist without the master invoice.

As the new system will be able to deal with 'off-the-street' customers an operational master of Customer Status is created to enable rapid identification of these customers.

Changes to supporting documentation

Any changes to the Logical Data Structure diagram will also result in changes to the supporting documentation. For instance, the new Driver Region and Invoice entities will require Entity Descriptions, modifications will have to be made to the Booking Entity Description, and new Relationship Descriptions will need to be created.

New requirements may also force changes to the supporting documentation without any changes to the diagram. Thus new items will need to be added to the Entity Description of Customer to enable the processing of 'off-the-street' Customers.

The Problems/Requirements List is one of the driving documents behind this step—as problems are solved and requirements met it should be amended to include solutions. Figure 4.14 showed how amendments for the required system processing were incorporated into the Problems/Requirements List, the incorporation of solutions from the required system data structure is exactly the same.

Validation against required system processing

On pages 92–7 ('Validating the Logical Data Structure') the validation of the current system processing against the Current System Logical Data Structure is described. The way in which the Required System Logical Data Structure is validated is exactly the same—it is not described here.

The Logical Data Structure should be validated against the update and enquiry processing. The update processing is detailed in the update Function Catalogue and on the Required System Data Flow Diagrams. The enquiry processing is detailed in the Retrievals Catalogue (sometimes called the enquiry Function Catalogue). Although some of the processing will be similar to that previously validated during the development of the current Logical Data Structure, an informal check should be made to ensure that all processing carried through into the new system can be supported by the new data structure. The new processing requirements should be thoroughly checked.

Entity/Data Store Cross Reference

The Entity/Data Store Cross Reference provides a further check that the data structure supports the processing. This cross-reference is first developed in the definition of the logical system (step 200 described in Sec. 3.2). Each current logical main data store contains current system entities. The contents of each data store are recorded on an Entity/Data Store Cross Reference form (see Fig. 4.3). This form should be extended during steps 240 and 250 to include the required system entities and data stores.

SUMMARY

The Current System Logical Data Structure and supporting documentation are extended to define the required system data structure.

New requirements which the Logical Data Structure must support are defined by the chosen Business System Option and the Problems/Requirements List.

The Problems/Requirements List is amended to describe any solutions adopted.

The Required System Logical Data Structure is validated against the required system processing and the Entity/Data Store Cross Reference is extended.

EXERCISE

Scapegoat Systems
Extend the Current System Logical Data Structure developed in the exercise in Sec. 3.4 by adding the new requirements:

1. For each employee store a list of all his or her qualifications including the subject and the level reached.
2. Be able to retrieve by subject and level, e.g. find all employees with an 'A' level or degree in French.

4.7   Sequencing changes to the data (step 250)

The use of Entity Life Histories in analysis

Entity Life Histories are a powerful analytical tool. The principal use of Entity Life Histories in stage 2 of SSADM is to find out how the new system should work. The technique is often considered to be the most challenging technique of SSADM as the drawing of a single Entity Life History may span several days. This aspect of the technique is not due to any particular difficulty inherent in the technique but is due to the abundance of questions raised that must be answered and the often heated, detailed discussions between members of the development team.

Entity Life Histories are not just a means of writing down previously discovered facts about a system. They are used as an investigative tool to find out more than is apparent in the Required System Data Flow Diagrams. Data Flow Diagrams do not show the interdependency of the processes within the system. Entity Life Histories explore in detail the context of each event that impacts on the system, showing clearly the permitted sequences of events that affect the data.

Section 2.4 described the relationship between Entity Life Histories and the other two main techniques of SSADM—Logical Data Structuring and Data Flow Diagrams. The use of Entity Life Histories to validate the other two techniques is summarized here:

• All Data Flow Diagram processes that update the main data stores are triggered by events. It is possible to identify events initially from the Data Flow Diagrams. Entity Life Histories are used to validate the Required System Data Flow Diagrams. When Data Flow Diagrams are large and complex, it is easy to omit necessary detail. An example of this would be data stores being used by processes without a process to create the data store in the first place. In examining the processing from the point of view of the data, the omission becomes obvious.

- Entity Life Histories are used to validate the Logical Data Structure. Difficulties in drawing an Entity Life History may be due to the poor identification of entities. If an entity contains repeating unrelated information, its Entity Life History may be very difficult to draw. This should stimulate a re-examination of the Logical Data Structure and new entities may be created.

Approach to Entity Life Histories

An Entity Life History charts all of the events that may affect an entity during its life within the system. An event is something that happens to trigger a process to update system data. The starting point for Entity Life Histories is the identification of events from the Data Flow Diagrams. The triggers to the processes that update main data stores are catalogued as events. Additional events are identified during the production of Entity Life Histories.

The events are associated with the entities they affect by using the Entity/Data Store Cross Reference. This association is documented on an Entity Life History (ELH) Matrix. After the Entity Life Histories are constructed with the addition of state indicators, the events are fully documented using the Event Catalogue.

The production of Entity Life Histories

The following sections describe the tasks involved in creating and documenting Entity Life Histories. In summary, these steps are:

- Initial identification of events.
- Constructing the ELH Matrix.
- Drawing initial Entity Life Histories for all entities.
- Completion of the Entity Life Histories.
- Addition of state indicators.
- Completion of the Event Catalogue.

Initial identification of events

Events are initially identified from the bottom-level Required System Data Flow Diagrams. As described above, an event is whatever triggers a process to update system data. On the Data Flow Diagrams, a process that is updating system data can be identified by the fact that there will be a data flow from the process box to a main data store (i.e. a data store which corresponds to one or more entities from the Logical Data Structure). By examining the process, it is possible to find the events by deciding what will cause the process to be set in action.

There are generally three types of event that can be identified in this way:

Externally sourced   The Data Flow Diagram will show a data flow from an external source/recipient to the process updating a data store (Fig. 4.20). This represents someone (or something) external to the system starting up a process because of something that has happened outside the system. An example of this would be the receipt of a booking request by an office of Yorkies. The office staff would wish to record this receipt on the system so would start up a process called 'Accept Booking'. The event that could be identified here would be 'Receipt of Booking Request'.

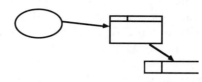

Fig. 4.20

Time based   This type of event will normally be identified by an update to a data store from a process that has no apparent triggers (Fig. 4.21). (Alternatively, it can be indicated by an input to a process from a 'Diary' or 'Actions' data store.) This represents a process being triggered at a particular time or on a particular date.

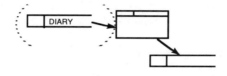

Fig. 4.21

Internally recognized   This type of event will be identified by an update to a data store from a process that has no apparent triggers in the same way as for the time-based event except that it will not be time or the calendar that is the trigger but some circumstance recognized by the system as a prerequisite to a process being triggered.

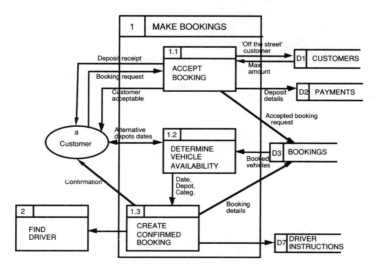

Fig. 4.22

Figure 4.22 shows the level 2 Data Flow Diagram of Process 1 from the Required System Data Flow Diagrams. The events are identified firstly by looking for updates (input data flows) to data stores. In this diagram, there are updates to the following data stores: D1, Customers; D2, Payments; D3, Bookings. (D7, Driver Instructions, is a manual data store and so does not map to system entities—the events affecting it need not be considered.)

What triggered these updates to be made? By tracing the flows back, firstly to the processes and then beyond, to the inputs to those processes, it is possible to deduce the events that must have triggered these updates.

D1 (Customers) and D2 (Payments) are updated by Process 1.1 (Accept Bookings). The input to this process comes (indirectly) from the customer. From the descriptions on the data flows, we can deduce that the event that causes these data stores to be updated is the receipt of a booking request. The Customer data store will only be updated if the booking request is received from an 'off-the-street' customer, in which case a new customer record will be created as the system will not have any previous record of this customer. (Booking requests received from customers already known to us will not cause an update to this data store.)

D3 (Bookings) is updated by Process 1.3 (Create Confirmed Booking). The event that triggers this update is 'Booking Confirmed' .

Constructing the ELH Matrix

As the events are identified, they are catalogued along one axis of the ELH Matrix. Entities from the Logical Data Structure (excluding operational masters) are placed along the other axis of the matrix. At the appropriate intersections on the matrix a letter is placed to denote whether the event causes the entity to be inserted (I), modified (M), or deleted (D). Sometimes the event may cause more than one of these effects, so more than one letter may be placed at a single intersection. If the event names chosen do not adequately describe the events, the Event Catalogue may be used to catalogue a description of each event to ensure a universally consistent view of what is meant by each event name. Figure 4.23 shows a portion of the Entity Life History Matrix for the Yorkies system.

Events / Entities	BOOKING REQUEST	NEW CUSTOMER	DRIVER ASSIGNED	AGENCY ALLOCATED	VEHICLE DEPARTURE	VEHICLE RELOCATION	VEHICLE RETURN	BOOKING CONFIRMED	PAYMENT DUE	PAYMENT RECEIVED	CUSTOMER CHANGE	NEW VEHICLE							
CUSTOMER	I	I									M								
INVOICE							I		M										
PAYMENT	I									I									
PAYMENT/INVOICE										I/M									
BOOKING	I		M	M	M	I	M	M											
VEHICLE							M					I							
VEHICLE CATEGORY																			
DRIVER																			
AGENCY																			

Fig. 4.23

After the initial drawing of the ELH Matrix, it is useful to review it before continuing to the Entity Life Histories. By reviewing the matrix from the point of view of the entities, it is useful to check that each entity has at least one 'I' against it because any data that is in the system must have been created by something. Similarly, by examining the matrix in the light of knowledge of the requirements, it is often possible to identify events that are missing. Events identified in this way are added to the matrix and effects on other entities explored. If it becomes obvious that there are some process boxes missing on the Data Flow Diagrams, then the Data Flow Diagrams should be amended to remain consistent with the Entity Life Histories.

Drawing initial Entity Life Histories
The Entity Life Histories are created in two passes:

→ initial (or 'simple') Entity Life Histories are created first;
→ final (or 'full') Entity Life Histories are produced afterwards.

This does not mean that two distinct sets of Entity Life Histories are produced in this step. The 'simple' Entity Life Histories are often quite complicated and the changes made to render them 'full' may be fairly trivial.

The first entities for which Entity Life Histories are drawn should be entities that have one or more masters but no details. Secondly, the masters of these entities are selected and so on until the entities at the 'top' of the Logical Data Structure are chosen. This is so that events that affect the relationships between entities are followed up through from detail to master. The effect of an event on each detail entity is considered first and then the effect from the other end of the relationship is viewed.

On the Yorkies Required System Logical Data Structure, the only entities without details are:

- Payment/Invoice Link
- Booking

Entity Life Histories are drawn for these entities first. From the ELH Matrix, we can see that the events that affect the Booking entity are:

- Booking Request (I)—this is where an initial booking request is received that is recorded on the system with a provisional status until it has been confirmed.
- Booking Confirmed (M)—this confirms a provisional booking after the vehicle has been allocated and all appropriate authorization received. At this point, it has not been determined whether a Yorkies driver is available or whether an agency driver will be required.
- Driver Allocated (M)—this is where a Yorkies driver is allocated to this booking.
- Agency Allocated (M)—this is where an agency is being used for the booking and has been assigned to the booking.
- Vehicle Departure (M)—the vehicle rental actually starts at this point; no further amendments can be made to it.
- Vehicle Relocation (I)—a vehicle from another area has been relocated to his area, bringing with it a booking. This sets up a Booking entity occurrence. At this point, a driver is not allocated as it must be a driver from this area.
- Vehicle Return (M)—this is when a hired vehicle returns to the depot and the details of the hiring are entered into the system. At this point, an invoice is produced.

The following paragraphs describe how the initial Booking Entity Life History is built up.

1. Firstly, the events that create (insert) a booking are identified. There are two Booking Request and Vehicle Relocation. This means that an occurrence of a Booking is created either by the receipt of a Booking Request or by the import of the bookings associated with a vehicle that has been relocated. The first thing we draw is a diagram as shown in Fig. 4.24. The entity name is in the box at the top to show which entity this life history belongs to. The box below that (Booking Creation) is called a 'node' as it has no significance within the Entity Life History except that it indicates a particular phase of the life history. The event names are placed in the boxes at the bottom of the structure. These boxes are called 'effects' because they represent the particular effects of the events upon this particular entity. The small circles in these boxes indicate a selection. (When drawing an Entity Life History, there would be no difference between the different boxes on the diagram. However, for this and subsequent diagrams here, the effect boxes are highlighted for clarity.) So, this first portion of the diagram shows that the Booking entity is created either by the Booking Request being received or by the relocation of a vehicle.

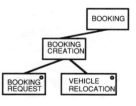

Fig. 4.24

2. Next, the events that will modify this entity are considered so that their appropriate position on the diagram can be determined. Are any of the events only going to happen

after one of the creation events, or are they going to happen after both of them? Are the modifications going to happen in any particular sequence, will some of them be alternatives for one another? Here, the Booking Confirmation will occur only after the Booking Request, not after Vehicle Relocation. The Driver or Agency Allocation will occur after either of them but will be alternatives to one another (we do not require both a Yorkies driver and an agency driver for a single booking but for some bookings we will need one and for other bookings we will need the other). The structure above needs to have these modifications added in the right place (see Fig. 4.25). There are several things to note here:

(a)   As Booking Confirmed can only happen after the Booking Request and not after Vehicle Relocation, the structure under the Booking Creation node has been amended to show a selection of: *either* the receipt of a request (shown as a node here) which consists of a sequence of the receipt of the request followed by its confirmation; *or* the Vehicle Relocation.

(b)   Effects must always be at the bottom of the structure which is why the Booking Request box has been moved to below the new node Receipt Of Request. It is not strictly necessary to label node boxes, but it is a useful indication of what is going on in that phase of the Entity Life History.

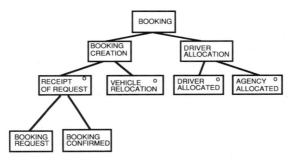

Fig. 4.25

(c)   The next thing that happens to the booking is that a driver is allocated, either from the Yorkies pool of drivers or from an agency. No other changes can be made to the booking before this allocation has been done.

If you are finding the diagrams difficult to understand, refer back to the explanation of Entity Life Histories in Sec. 2.3.

3. Referring back to the ELH Matrix, there are just two more events that affect this entity: Vehicle Departure and Vehicle Return.

From our knowledge of the system, it is obvious that the driver must be allocated before the vehicle departs and that the vehicle must depart before it returns! Here is a simple sequence, then, that is reflected on the Entity Life History as shown in Fig. 4.26.

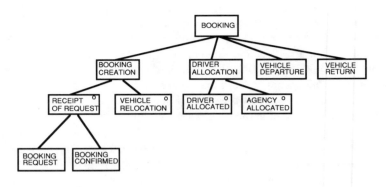

Fig. 4.26

4. Now, it is necessary to stop using the ELH Matrix as there are no more events indicated there that affect the Booking entity. We need to decide whether or not there are events that affect this entity that have been missed. The Entity Description is used to check that each data item is created, possibly modified and deleted. The Entity Description of the Booking entity is shown in Fig. 4.27.

Entity Description

Entity   : Booking
Description   : An agreement made between a Yorkies Ltd Local Office and a customer to
                supply a vehicle or vehicles of an agreed category and possibly drivers on a
                particular date or series of dates.

System : Yorkies	Sub-system :		Author :	Date :

Key	Data items	Format	Length	Comments
/	Booking No.			
*	Customer No.			
*	Vehicle Category Requested			
	Date Booking Received			
	Date Booking Confirmed			
	Date Booking Starts			
	Date Booking Ends			
	Driver Required/Not Required			
*	Vehicle Registration Mark			
*	Driver No./ Agency No.			
	Start Mileage			
	Finish Mileage			
	Date Collected			
	Time Collected			
	Date Returned			
	Time Returned			
*	Office No. (start)			
*	Office No. (finish)			
*	Invoice No.			

/,Prime key; *, Foreign key.

Fig. 4.27

Looking through this list, the following data items would be present when the booking is created:

• Booking No.;
• Customer No.;

- Vehicle Category Requested;
- Date Booking Received;
- Date Booking Starts;
- Office No.(start);
- Office No.(finish);
- Date Booking Ends;
- Driver Required/Not Required.

The Booking No. is the key of this entity, and therefore needs to be present for a unique occurrence of the entity to be created. The other data items are either part of the booking request (i.e. Customer No., Dates start and finish, Vehicle Category Requested and Driver Required/Not Required) or recorded at the time of registration (i.e. Date Booking Received, Office No.(start), Office No.(finish)).

The items Date Booking Confirmed and Vehicle Registration Mark are added when the booking is confirmed.

All the items above are present when a vehicle is relocated to another office so would be created by the Vehicle Relocation event.

Driver No./Agency No. is updated when a driver has been allocated or, if a Yorkies driver is unavailable, an agency has agreed to provide a driver.

The Vehicle Departure would update the Start Mileage and Date and Time Collected and the Vehicle Return would update the Date and Time Returned and Finish Mileage.

This quick check has shown us that most of the data items are created by one or more of the events reflected in the Entity Life History. The one data item that is not affected by any of these is Invoice No. This is a foreign key, i.e. it is the key of the Invoice entity which is the master of Booking. The reason for the presence of this data item in the Booking entity is to represent the relationship between the two entities. This means that when the invoice is issued, a relationship between invoice and booking is created, so an update to the Booking entity should be shown. The invoice will only be issued after the vehicle has been returned, so the Entity Life History is amended as shown in Fig. 4.28.

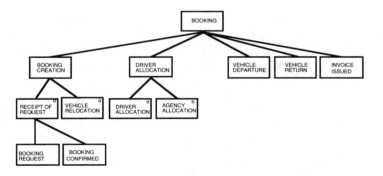

Fig. 4.28

5. Now, the data items should be examined again: are any of these items likely to change after it has been created? (At this point, only the normal things that will cause the items to change should be identified: the more unusual or random events will be identified in the

next pass of the Entity Life Histories.) As people are often known to change their minds, it seems likely that the customer may wish to alter details of the booking once it has been made. For example, the start and end dates may change, or the driver requirements may change. If the load to be carried increases in size, a different vehicle category may be required. All these changes could take place a number of times before the vehicle actually leaves on hire. This means that the Entity Life History must be changed again to show that amendments can be made to the booking details before the vehicle departure. However, it will be necessary to confirm the booking again after any change has been made. This results in the Entity Life History shown in Fig. 4.29. Here, an iteration of amendments has been added (shown by an asterisk in the top left corner of the box). The repeated element is a sequence of a booking amendment followed by a confirmation. This means that the event Booking Confirmed now has two effects upon the Booking entity: the first is immediately after the booking has been created by a booking request and the second immediately after a booking amendment. This will often happen in the construction of Entity Life Histories as it is quite normal for an event to affect an entity at several different points in its life.

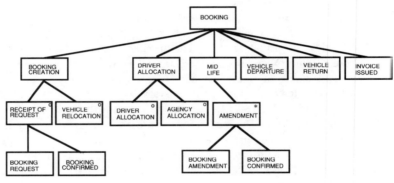

Fig. 4.29

6. The next question to ask is 'How would the entity normally be deleted from the system?' It will not be possible to delete the record of the booking immediately after an invoice has been issued as there may be queries to resolve with the customer before he pays the invoice. Also, it might be useful to refer to old bookings when a customer makes a new booking. Even after the record is removed from the live system, there may be a requirement to keep the booking information for future reference. This means that the event that finishes this Entity Life History is 'Archive'. This is because this life history is only interested in the life of the entity while it is on the live system's database. After archive, the Booking entity will be held on a tape for some time. The final 'simple' Entity Life History for the Booking entity now looks like Fig. 4.30.

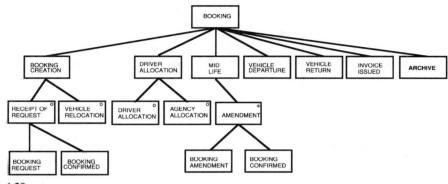

Fig. 4.30

The example developed here for the Yorkies Booking entity illustrates some important points about developing the initial Entity Life Histories. The points, together with some additional hints are summarized here.

Entity Life Histories should be made as unambiguous as possible   The diagram should not be open to several different interpretations. We could have drawn the first part of Fig. 4.30 in the form shown in Fig. 4.31. (The 'null box' under the selection shows that either nothing further will affect the entity after the Booking Creation phase until Driver Allocation or that the Booking Confirmed event will affect the entity. This is a clumsy way of showing that Vehicle Relocation is not followed by a Booking Confirmation)

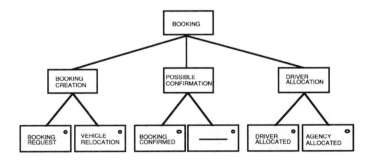

Fig. 4.31

Figure 4.31 will still allow the same valid sequences to occur:

- Booking Request, Booking Confirmed, Driver Allocation;
- Vehicle Relocation, Driver Allocation.

However, ambiguity has now been introduced as other sequences, which are not valid, are now possible:

- Booking Request, Driver Allocation;
- Vehicle Relocation, Booking Confirmed, Driver Allocation.

So, although the diagram may be said to reflect the required sequences, the diagram is incorrect because ambiguity about valid and invalid sequences has been introduced.

Selections under iterations should be examined   If an Entity Life History contains a selection under an iteration, i.e. a selection that may be made any number of times, each box under the selection should be examined to ensure that this really reflects the requirement. It is possible that some may occur only once, or that some may happen in sequence. It is important to ensure that these are not hidden in an iterated selection.

Only constraints imposed by this system should be reflected   It is tempting to reflect the constraints and sequences of the real-world system in the Entity Life Histories. However, it is important to reflect the requirements stated by the users rather than allow the Entity Life Histories to become too constraining. For example, a pay system will be interested if a member of staff is promoted as he or she will need to be paid more. However, the exact progression of promotions, and all associated factors, is of no interest to the pay system. So, although it would be possible to add a quite complex structure of promotions to a 'Person' Entity Life History within the pay system, it is necessary only to show an iteration of 'Change In Rank' as that is all there is of interest to this system (see Fig. 4.32).

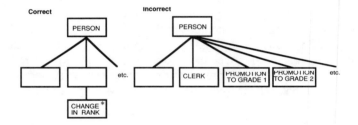

Fig. 4.32

Check that all events have been identified   The following list can be used to check that all of the events have been identified in the simple Entity Life Histories:

- Are all of the data items in the entity given values when the entity is created? If not, what events will give values to them?
- Is it a requirement that individual data items are updated, and what events will cause them to be updated?
- What will cause the entity to change its relationships with occurrences of the master entities?
- What causes an optional relationship to a master entity to be created or deleted?
- Is it important to know that an event has happened in the life of an entity even if there is not an obvious change in the value of any of the data items? It may be that certain events set the context for other events, updating only a flag or indicator in the entity.

It is necessary to try to account for all of the events that will normally affect the entity during its life in the system. However, the more unusual events that may occur should be left to the second pass of the Entity Life Histories. Whatever is expected to happen should be reflected in the simple Entity Life Histories.

Complete the Entity Life Histories
This time, the Entity Life Histories are dealt with working from the top of the Logical Data Structure to the bottom. This is called the 'downward pass'. This time, the interdependence of Entity Life Histories can be looked at and any unusual events can be included.

Interdependencies The effect of the deletion of each master entity on its detail entities is investigated next. For each pair of master and detail entities, only one of the following situations will apply. The deletion of the master:
- has no effect on the detail;
- causes the deletion of the detail;
- cannot occur until the detail has been deleted.

If the deletion of the master does not affect the detail (this will normally be where the relationship between them is optional), then neither Entity Life History is affected. An example of this would be if a driver were to leave Yorkies, the Driver entity could be deleted without affecting the Booking entity. This is because the existence of the booking does not depend upon the existence of a driver.

If the deletion of the master causes the detail to be deleted, then any events that are shown at the end of the Entity Life History of the master should be put at the end of the life of the detail. An example of this in the Yorkies system would be if a customer is deleted, the Booking occurrences belonging to that customer would also be deleted. This means that the Booking Entity Life History would now look like Fig. 4.33. A Booking occurrence will be deleted either by the normal Booking Archive for an existing customer or it will be deleted when a Customer entity is archived.

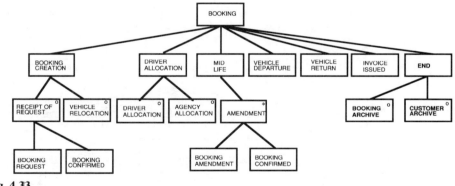

Fig. 4.33

If the master cannot be deleted until the detail has been deleted, then the events that cause the deletion of the detail should be put before the end of the master's Entity Life History. In Yorkies, a Vehicle occurrence may not be deleted from the system until all bookings belonging to that vehicle have been deleted. The end of the Vehicle Entity Life History might now look like Fig. 4.34. This shows that the last booking for this vehicle must have been deleted before the vehicle can be deleted from the system.

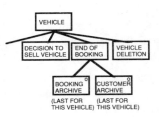

Fig. 4.34

Similar interdependencies may occur within the Entity Life Histories concerned as well as at the end.

Unusual events Events that occur unusually that alter the sequence of an Entity Life History are added now. We have assumed that the normal course of events is for a customer to make a booking and for that booking to go ahead as planned (the vehicle leaves at the beginning of the hire period and returns at the end of that period). There are quite a few things that might upset this, however. For example, the customer might decide to cancel the booking at any time before the vehicle departs. Also, the vehicle might be involved in an accident and never return. These events, and any other similar events, should be added to the Entity Life History at this stage. The Booking Entity Life History would be expanded as shown in Fig. 4.35. This structure shows that a booking can be cancelled after the initial set-up and before the vehicle departs. If the booking is cancelled, the next thing that will happen is the archive of the booking. If the vehicle departs, it will either return or it will be written off. Obviously, there are many more complexities that could be added (e.g. what happens if the customer dies or wishes to extend the period of the booking after the vehicle departs?) but we shall assume that this completes the Booking Entity Life History.

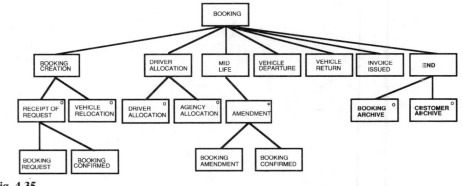

Fig. 4.35

It is worth noting that the Booking Entity Life History is likely to be the most complex Entity Life History within the Yorkies example as the Booking entity is the most 'active' within the Yorkies system. It is our experience that the majority of Logical Data Structures have one entity (or a very small group of entities) that is pivotal to the system. These entities are identified by the fact that they are central to the purpose of the system. For example, the Yorkies system is to help track bookings, therefore the Booking entity is central to the Logical Data Structure. Other examples might include personnel systems where the Person entity is central, banking systems where the Account entity is central, order processing systems where the Order entity is central, etc. Thus the Booking entity has been chosen to demonstrate several points about Entity Life Histories. The majority of the other Entity Life Histories will be very simple indeed. Examples are given in Figs 4.36–4.38.

In Fig. 4.36 the customer is first notified to our system either by applying to become a regular customer, or by requesting a booking as an 'off-the-street' customer. The only amendments that will be made during the customer's life within the system will be either a basic change to the customer's details (e.g. the address for invoicing, credit limit, etc.) or will be to do with the total amount owed, i.e. issuing of an invoice and receipt of a payment. At some time the customer may wish to withdraw from custom with Yorkies, after which the customer is archived.

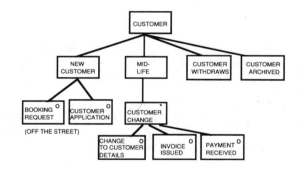

Fig. 4.36

In Fig. 4.37 an agency is notified to us, there may be a number of changes in details, such as address, during its life in our system, and then it might close. Alternatively, we may decide that we do not wish to do business with the agency any more and we delete it from the system.

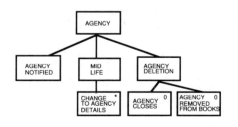

Fig. 4.37

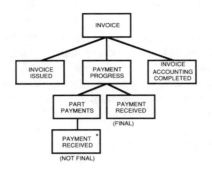

Fig. 4.38

The Invoice Entity Life History (Fig. 4.38) illustrates how an event can appear more than once in a single Entity Life History because it has different effects. Here, the Payment Received event is shown as an iteration with the qualifier Not Final followed in sequence by a single Payment Received (Final). This shows that while the invoice remains not fully paid, an iteration of a series of payments is allowed. As soon as the invoice is paid in full, the iteration finishes and the next event to affect this entity is the Invoice Accounting Completed which deletes the entity. (Just to remind you, if the Invoice is paid in full by a single payment, this structure is still valid as the iteration convention allows 'zero' occurrences as well as many.)

Addition of state indicators
A state indicator can be thought of as a data item within an entity that is updated every time an event affects it. This means that the value of the state indicator will show where in its Entity Life History an entity is at any one time.

The values given to the state indicator have no significance provided that each effect assigns a unique value to it. The convention is to start with a value of 1 and increment it by 1 each time an event affects the entity.

The first thing to do is to assign a value to each of the effect boxes on the Entity Life History as shown in Fig. 4.39. The numbers underneath the effect boxes show the value that the state indicator has been set to after the event has finished affecting the entity occurrence. For example, after the occurrence has been created by the Booking Request event, the state indicator is set to the value '1'. Therefore, if we find that the state indicator of a Booking occurrence is '1', we know that it has just been created but has not yet been confirmed.

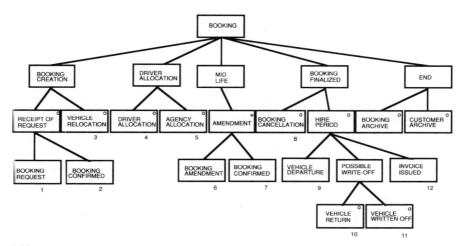

Fig. 4.39

Immediately it becomes obvious that the state indicator values can be used to help in determining whether or not we can allow an event to affect an entity. If the state indicator is not equal to '1', the Booking Confirmed event cannot be allowed to affect this entity. So, for each effect box on the Entity Life History, we can define a set of values of the state indicator that are valid before the event can be allowed to affect the entity. These are added to the diagram in front of the 'set to' value, separated by a '/'. If there is more than one 'valid previous' value, these are separated by commas. So, if the 'set to' value of the state indicator is '4' and the 'valid previous' values are either '2' or '3', then the set of values shown underneath the box will be 2,3/4. The Booking Entity Life History now looks like Fig. 4.40.

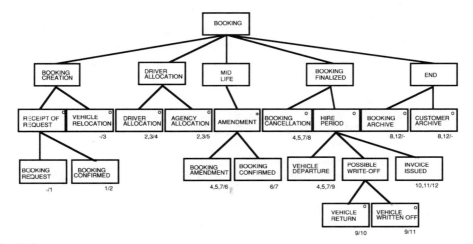

Fig. 4.40

The state indicator values depend entirely upon the structure of the Entity Life History:

- Creation effects have a null valid previous value because the state indicator does not exist before the creation event.
- Where there is a selection, all effects that are alternatives to one another have the same set of valid previous state indicator values as shown by the Driver Allocation and Agency Allocation both having the valid previous values of 2 and 3.
- Where there is an iteration, the value set by the repeating effect will also be included as one of the valid previous values for that effect. This is shown by the fact that one of the valid previous values for the Booking Amendment effect is '7' which is the value set by the subsequent Booking Confirmed effect.

The Booking Entity Life History does not contain either quits and resumes or a parallel structure (see Sec. 2.3). If these conventions are used, the state indicators are assigned as follows:

- Where the Quit and Resume is used, the value set by the effect with the 'Q' is one of the valid previous values for the effect with the 'R' and cannot appear as a valid previous value anywhere else on the diagram.

The rules for parallel structures are more complex. Each leg of a parallel structure is independent of the others, so one state indicator is not adequate to track the progress of the Entity Life History. There are two ways of dealing with this.

Either only one leg updates the state indicator In some circumstances, it is necessary only to follow one of the legs using the state indicator: the other legs are unimportant in terms of tracking a particular entity through its Entity Life History. To show that a state indicator remains unchanged after a particular effect, an asterisk is put in the place of the 'set to' value.

Or a separate subsidiary state indicator is introduced for each leg after the first Each subsidiary state indicator may be thought of as an additional data item in the entity and is updated by one leg of a parallel structure. Possible values of other state indicators may be shown as valid previous values for the leg that uses a subsidiary state indicator but only one state indicator is updated by each leg The terminating state indicator values from all legs of a parallel structure must be shown as valid previous values for the effect(s) that follow the parallel structure.

Completing the Event Catalogue

The Event Catalogue may be started at the same time as the Entity Life History Matrix to establish the definitions of all events that are identified. Each event is catalogued under the Data Flow Diagram process it triggers (or the function it is associated with if a Function Catalogue is available). This will help to identify processes that are missing from the Data Flow Diagrams by ensuring that somewhere in the Data Flow Diagrams there is a process that is triggered by each event. It is important to keep the Data Flow Diagrams up-to-date as they form part of the final system specification.

As the Entity Life Histories for the Yorkies system have been developed, the new events that have been identified have been added to the ELH Matrix. The matrix for the Booking entity now looks like Fig 4.41.

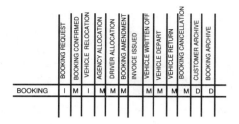

Fig. 4.41

Each of these events should now be catalogued in more detail. The form used to catalogue the events repeats the information contained in the ELH Matrix and adds a description of the events and the volumes expected for each event. This information helps in estimating how powerful the new system must be to support all of the events.

An example of the Yorkies Event Catalogue is shown at Fig. 4.42. The events are given an identifier to facilitate cross-referencing with other documents such as the Logical Dialogue Outlines and Logical Update Process Outlines. If there is a peak in the volume of certain events during the day, this should be indicated in the 'Comments' column. For example, in the Yorkies system, many of the bookings for vehicles are received by post, so the volume of the event 'New Booking' will peak twice in the day after the arrival of first and second post.

If there are sub-events identified within the system, then these should be grouped together on the Event Catalogue and the description column should include the criteria for the selection of a particular sub-event after the arrival of a single trigger.

Event Catalogue

System : Yorkies	Sub-system :		Author :		Date :			
ID	Events	Description	Entities affected	Effect (I,M,D)	Volumes			
					Ave	Max	Per	Comments
e1	Booking Request	A request for a booking is received	Booking Customer	I I	1400	2000	Day	Peaks at 10.00 a.m. and 2.00 p.m. postal deliveries
e2	Booking Confirmed							
e3	Vehicle Relocation							
e4	Agency Allocation							
e5	Driver Allocation							

Fig. 4.42

SUMMARY

The steps in the production of Entity Life Histories are:

1. Create the Entity Life History Matrix and optionally describe the events on the Event Catalogue form.
2. Draw simple Entity Life Histories, working from the bottom of the Logical Data Structure to the top, identifying all events that normally affect the entities.
3. Complete the Entity Life Histories working from the top of the Logical Data Structure to the bottom, establishing Entity Life History interdependencies and identifying events that would abnormally affect the entities.
4. Add state indicators to all Entity Life Histories.
5. Complete the Event Catalogue adding the volumes of the events.

EXERCISE

Scapegoat Systems

The Project entity was identified during the Scapegoat Logical Data Structure exercise in Sec. 3.4. Use the suggested answer given in Appendix D.

The following events can affect an occurrence of the Project entity.

A Project is initiated by either a user request (for an internal project) or by negotiation with a customer. When either internal agreement has been reached or a contract has been signed with the customer then the project will be entered onto the system.

A project manager is selected and then staff are allocated to the project. The project manager develops project plans which are entered onto the system when they have been agreed. A project will be made up of several distinct stages: the start date and end date of each stage are entered into the system as they occur. After each stage is completed the work undergoes a quality assurance review.

On completion of the whole project a full review is carried out and the results recorded on the system. One year after the final review the project is archived.

Develop an Entity Life History for the Project entity

4.8   Prototyping and dialogue specification (step 250)

Introduction

Batch and on-line systems

Most systems being developed today have a large proportion of on-line processing. This means that the user has direct access to the system via a terminal. Unlike traditional batch systems, where predefined information is entered in large batches to be processed in a predefined way, an on-line system offers much more flexibility, allowing the user to dictate the structure of a session at a terminal. To underline the difference between the two types of processing, imagine that you want to communicate some information to a friend. This information is sent as:

- A *batch*, if you enclose all of the information in a letter. You perhaps plan the letter, decide what you want to say, then write it and mail it. Once the letter has been sent, there is no possibility of changing the information in there. If you do want to amend the information in that letter, you will have to send a second letter contradicting the first. If your spelling is wrong, or the address is incomplete, the postal service and your friend will have to cope, based upon their knowledge of geography and spelling.

- A *dialogue*, if you phone your friend. No planning is required—you dial the number, and blurt out all your news as you think about it. If there is anything that you friend does not understand, you will be asked to repeat or rephrase your news. During the conversation, your friend may give you information that causes you to modify some of the news you were about to impart.

In a similar way, a batch system is relatively straightforward when designing the input stream of data, but the error processing must be built in to cope with error data. This can be a very complex part of the design of batch processing. An on-line system is less straightforward when designing the facilities to input data. The user might want a combination of ease of use and sophisticated control over the course of dialogues with the system. The error processing is much easier than in batch systems. As the user is sitting at the terminal when the data is first input, if the system detects errors, it can ask the user to rekey the information until it is correct.

Importance of on-line systems

Most computerized information systems under development will have mainly on-line processing. In the past batch systems were cheaper to build and use, but current hardware and software costs mean that there is now little cost difference between batch and on-line processing. As on-line processing offers more accurate and timely information it is now generally favoured.

On-line processing has a much greater impact on its users than batch processing. Users interact directly with the computer in their work-place. This means that the interface will largely determine the user's reaction to the new system. Thus the design of the interface is critical in ensuring the success of a new system. A user will have a view of a particular task; what information is required to perform it, the tasks that precede and succeed it, and of the ways in which they like to perform this task. A computerized information system will impose a view of the task upon the user. For a successful implementation of a task the user's view and the system's view must be as close as possible.

The place of dialogue design in SSADM

SSADM uses a number of techniques to try and make the view of the users and the view of the system developers as close as possible. So far we have discussed techniques for modelling: the hierarchy of activities and the information that flows between them (Data Flow Diagrams), the information that the system should hold (Logical Data Structures), and the dynamic interaction of events with the system's data (Entity Life Histories). We have not discussed any techniques for modelling or specifying the users' dynamic interaction with the system in on-line processing.

The earliest point that modelling of the interface can start is when the functions which are to be performed interactively in the new system have been agreed. In SSADM the basic unit of processing is the event which was identified in the creation of Entity Life Histories, events will later be used to specify the processing in more detail. For each event to be handled on-line there will be a dialogue between the user and the computer. It is this dialogue which is modelled initially in SSADM stage 2.

Dialogue design is often divided into three main components: content, mode of control, and format. Content refers to the data and messages passing between the user and the system. Mode of control refers to the way in which the user moves from one dialogue to another. Format refers to the layout of data and messages on the screen.

To a large extent the format of screens will be influenced by the hardware and software available: factors such as colour, screen size, keyboard layout, windowing facilities, and many others. The mode of control is less dependent on the physical environment although some fourth-generation software may impose limitations on menu structures. The content should be completely implementation independent—it should be governed by the underlying business requirements.

It should, then, be possible to specify the content of a dialogue at the end of stage 2 without either prejudicing the physical design or needing to repeat the content specification when the physical environment is known. Design of the mode of control and particularly format design should be left until the physical environment is decided in stage 3.

In SSADM a technique, logical dialogue design, is used to specify the interaction between a user and the computer for each event to be processed on-line. It is performed in step 250, *investigate detailed system logic,* after the on-line events have been identified. This technique concentrates on the content and the actions performed by the user rather than the format of the screens. Logical dialogue design is described in the next section.

Figure 4.43 shows how logical dialogue design fits into the stages of SSADM. In stage 5 the Dialogue Outlines are used to help produce the Logical Update Process Outlines. Dialogue Outlines produced in stage 2 are cross-referred to the Process Outlines and Dialogue Controls are produced to link the event level dialogues together. In stage 6 the logical dialogue design is used as input to first cut program specification; critical on-line processing will then be timed during physical design control. When a tuned design has been produced the Dialogue Outline and Controls will form an important part of on-line program specification.

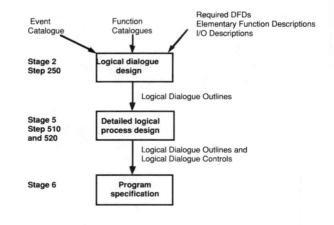

Fig. 4.43

Many SSADM projects are now using prototyping techniques to help specify dialogues. This alternative or addition to logical dialogue design is discussed on pages 157–9, 'The use of prototypes'.

Logical dialogue design

Purpose

Logical dialogue design specifies the content and the actions performed by the user and by the computer required for the processing of an event which is to be processed on-line. This dialogue design is used in two ways in SSADM: to ensure that the dialogues meet the requirements of the users and as specifications for their physical implementation.

Users find the way in which dialogues are represented in SSADM easy to understand. They can see from the logical dialogue design exactly what information they will directly input into the computer, what information the computer will display to them, the actions they will perform, and, at a high level, the actions the computer will perform. The specification of the on-line processing can therefore be easily checked by the user before detailed logical and physical design of the processing begin.

In order to design the physical screens and detailed physical interfaces, system designers and programmers need specifications of the information content of dialogues.

Notation

A Logical Dialogue Outline is developed to represent a dialogue for each on-line event. The outline comprises a diagram showing the mode of control, details of the data items input and output by the computer, the decisions made by the user and by the computer, and a broad description of the processing performed by the computer. The symbols used for the diagram are similar to those used in conventional flowcharts (see Fig. 4.44). Some of the components shown in Fig. 4.44 require further explanation.

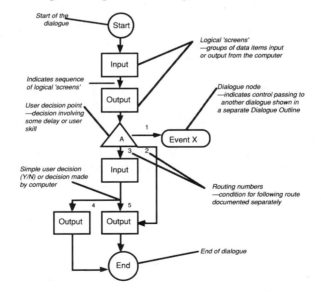

Fig. 4.44

Logical screens These can be thought of as parcels of information passing between the user and the the computer. The way these logical screens are physically implemented will depend on the software and hardware chosen. It may be possible to implement the whole dialogue with one physical screen or it may be implemented with several physical screens per logical screen—this can only be determined when the physical environment has been decided. In physical design (stage 6) references to physical screens are added to the logical screens on the dialogue outline.

Dialogue nodes These represent the passing of control to another dialogue. They will result from the computer or the user making a decision that leads to the processing for another on-line event—this will be represented by a separate Logical Dialogue Outline. In stage 2 dialogues are represented separately for each event. In stage 5 (Detailed logical process design) dialogues are linked together to form Logical Dialogue Controls. These use the same notation to describe the ways in which the event level dialogues are reached.

User decision points These are shown by triangles on the diagram with a letter inside them cross-referring to a description. A user decision will involve the user in performing some actions in order to make this decision, this could be on the basis of other manually held information or making a value judgement based upon the information presented. It will normally require some skill and delay to make the decision. The possible routes which depend upon the decision are identified with routing numbers so that the conditions for each route can be described.

Simple user decision points and computer decisions These are both shown by the splitting of the routing lines into two or more directions. A simple user decision point is one where the user can make yes/no decisions, e.g. repeating the dialogue or requesting more information. A computer (or system) decision is one where the program decides which route should be followed based upon the information available.

Figure 4.45 shows a Logical Dialogue Outline form in which the diagram is annotated to show the data items being input and output, the processing performed by the computer, and the conditions applying to each decision point.

   The data items column shows the data items input into and output from the computer. Sometimes all items associated with the entity might be required—the entity name is then used. Sometimes precise items have not yet been agreed and a summary description is given. Messages sent by the computer e.g. 'no records found', are sometimes recorded in this column. Although the diagrams should not be complicated by showing error processing, validation, and editing, they should show the major paths through the dialogue. The large arrows show the direction of information flow between the user and the computer.

The processing comments column should give an outline description, in terms the user can understand, of the processing the computer will perform to produce its next output. Simple user decisions and system decisions are recorded in this column together with the conditions applied to decide which routing will be followed. In stage 5 the Dialogue Outline is cross-referenced to the Logical Process Outline by including the Operation ID for the relevant processing on the Dialogue Outline. The Process Outline will describe the processing in much greater detail.

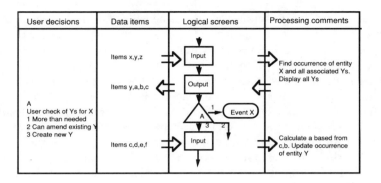

User decisions	Data items	Logical screens	Processing comments

Fig. 4.45

The user decision column is where the conditions and the processing applied by the user are described. In Fig. 4.45 user decision 'A' is described in this column with the conditions under which routes 1, 2, and 3 are followed. One of the strengths of this technique is that users can immediately see which decisions are to be made by them and which are to be automated.

Method of development

This chapter is mainly about logical dialogue design in step 250, *investigate detailed system logic*, so the way in which Dialogue Outlines are developed will be discussed. A systematic or 'cookbook' approach to developing the outlines is described—in practice analysts will develop their own approach to this activity. This suggested approach is shown diagrammatically in Fig. 4.46.

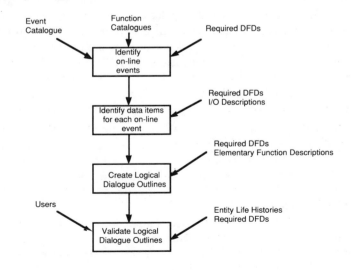

Fig. 4.46

Each of these tasks is now discussed using examples from the Yorkies case study.

After completion of the Entity Life Histories all events should be described in the Event Catalogue. The Function Catalogue, developed with the Required System Data Flow Diagrams, is split into on-line and batch sections. In order to decide which events are to be processed on-line it is necessary to match the events to the on-line functions. This should be straightforward—if there is any problem in assigning an event to a function then the required Data Flow Diagram should help. Each event should result in an update to a data store and so the event should be handled by a process, this process will be assigned to a function—in this way the required Data Flow Diagram can help identify the on-line events. The process handling the event is sometimes identified on the Event Catalogue.

It may not be worthwhile to create a Dialogue Outline for every on-line event; the processing for some may be trivial, e.g. find occurrence, display, and make modifications. Dialogue Outlines should only be created for the events requiring complex on-line processing, a good rule of thumb is to develop an outline for any requiring more than four logical screens.

A good approach is to go through the Event Catalogue with the on-line Function Catalogue and identify the events which will require Dialogue Outlines. The Yorkies Event Catalogue is shown in Fig. 4.40 and the on-line Function Catalogue is shown in Fig. 4.15.

Identify on-line events The event Booking Request is handled by on-line Function 4, Make Booking. This will require a Logical Dialogue Outline. Other events requiring Dialogue Outlines are identified in a similar way. The development of a Logical Dialogue Outline for the event Booking Request is described below.

Identify the data items The data item content of the dialogue needs to be decided. The process handling the event will have data flows in and out. Data flows crossing the system boundary will have been documented by I/O Descriptions detailing their data contents. Some projects document the contents of all data flows in a data dictionary. These descriptions of data content will help decide which data items are involved in the dialogue.

The event Booking Request is handled by Process 1.1 Accept Booking. The data flow from customer to Process 1.1 is documented by the I/O Description 'a-1.1' shown in Fig. 4.17. Other information about the data items required for this dialogue can be found from the data flow names on the level 2 Data Flow Diagram for Process 1 Make Bookings shown in Fig. 4.11. Note that not all of the items discovered are relevant to the particular dialogue since the Data Flow Diagram is handling several events.

Create Logical Dialogue Outline The data items involved in the dialogue have been determined. These now need to be assigned to logical screens and the structure of the dialogue built. This task is also based on previous SSADM products particularly the lowest level Required System Data Flow Diagrams and their supporting Elementary Function Descriptions. The relevant Yorkies Elementary Function Descriptions were shown previously in Fig. 4.14.

The Dialogue Outline will be based on previous SSADM products, but, unless these are very detailed, will not be a reformulation of them. Designer skill is involved to interpret what the users really require and to ensure that the dialogue can be computerized (although not yet concerned with physical factors, the eventual mapping to software must be possible). The development of Dialogue Outlines is iterative; they are drawn, discussed with users and redrawn until agreement is reached. Figure 4.47 shows the Logical Dialogue Outline for the event Booking Request.

Validate Logical Dialogue Outlines Each Logical Dialogue Outline is validated against the Entity Life Histories to ensure that the processing resulting from the event occurs at the right time in the life of the entity. The Dialogue Outline is also reviewed against the Required System Data Flow Diagrams to ensure that all on-line processes and data flows have been considered.

Most important of all is validation with the users. They find it easy to understand Logical Dialogue Outlines and should gain a clear view of how the on-line processing will affect them. Users should review: the data items input and output, the decisions to be automated or under manual control, the sequence of screens through the dialogue, and the computer processing. Several iterations of design and discussion may be necessary for each dialogue.

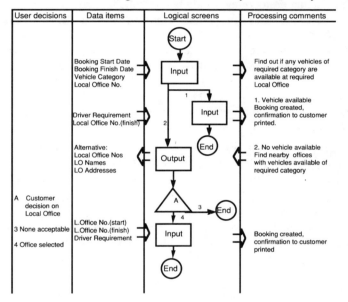

Fig. 4.47

The use of prototypes

Introduction

Prototyping has been described as an approach 'lacking tightly written systems design specifications' which is capable of 'providing the user with a tentative system for experimental purposes at the earliest possible time' and which can evolve into the production system. Prototyping in information systems design has been subject to much discussion recently, with many articles and books being published. There are two major reasons for this recent interest:

Users find it impossible to specify their requirements

The future users of a system find it very difficult to imagine what it will look like until it is in front of them. It is very common for a project team to feel sure that a thorough analysis has been performed, but after implementation many enhancements are requested by the users. Only by seeing what the system looks like and experimenting with some of the facilities can users decide what is really required.

Modern tools enable rapid systems development

Many projects now use fourth-generation software for systems development. This software provides a number of facilities that make rapid construction of business information systems possible. They provide some or all of the following facilities: relational database management systems, data dictionaries, very high-level languages, end-user query languages, report writers, tools for building interactive systems such as screen painters, and tools for analysis and design and sometimes generation of code from these.

Although prototypes are built early in the systems development life cycle, most authors agree that some basic data and function modelling are necessary first. It is also widely accepted that prototyping should be controlled as part of a systems development method.

There are basically two kinds of prototype: ones which will be thrown away and ones which will eventually be developed into a full production system. The 'throw away' variety are used as a learning medium between the developer and the user to help them converge on an adequate set of requirements. Typically 'throw away' prototypes are built very rapidly, very early in requirements analysis but will sacrifice certain aspects of the system such as error handling, security, and performance. The 'developable' prototype will evolve into a working system; sometimes 'throw away' prototypes turn into 'developable' ones.

Prototyping in SSADM

'Fourth-generation' tools are being widely used to construct information systems that have been analysed and designed using SSADM. Much work has gone into defining interfaces between SSADM and a variety of products. Fourth-generation tools are also being used to develop prototypes in the early stages of SSADM projects and variations of SSADM have been developed which incorporate prototyping.

Three variants of prototyping are defined for SSADM: levels 1, 2, and 3. These are integrated into the steps of the method and can be integrated together so that a level 1 prototype may evolve into a level 2 which can then evolve into a level 3.

Screen-based dialogue simulation (level 1)

A throw away prototype is developed of the on-line dialogues. There will be no underlying database or code for storage or manipulation of the data input. The prototype is usually developed in step 250 in addition to or as replacement for the Logical Dialogue Outlines.

Some of the microcomputer database packages provide very convenient facilities for dialogue simulation, such as screen painters and menu handlers. The I/O Descriptions for some of the major events and enquiries can be turned into screens using a screen painter. In its simplest form the analyst can lead the user manually through a series of screens, explaining how each is used, to simulate a dialogue. To provide a more realistic simulation the screens could be linked together and high-level menu structures built.

The advantage of the simulation is that the user can get a good idea of the facilities and how the dialogues will look at the early stages of a project. It also, like all prototyping, encourages user participation and increases user satisfaction.

The disadvantages are that the simulation will not be very realistic and some imagination will be required for successful user evaluation. Another potential disadvantage is that the evaluation can concentrate too much on the format of the screens rather than their content or the mode of control. The hardware and software environment are often not decided until after stage 3, so detailed evaluation of screen formats could prove fruitless.

A throw away, working prototype (level 2)

This is an extension of the level 1 prototype in that an underlying database structure and some data manipulation code are produced. It then becomes a working prototype implementing some of the system's important facilities. It is developed in the detailed logical process and data design stages of SSADM (stages 4 and 5).

This form of prototype is most useful when it draws upon the technical option selected in stage 3. The prototype can then be built using similar tools to those to be used for the final construction. This means that the style of user interface will be closer to the final implementation thereby allowing better evaluation. It also means that the prototype will be less throwaway in that data dictionary screen and data definitions may be preserved for final construction.

A working prototype (level 3)

Here a sub-system area or set of functions are taken and developed into a working prototype. This will then form the nucleus of the new system with the remaining functions and sub-systems later developed around it to form the full production system. Level 3 prototypes are developed in the physical design stage of SSADM (stage 6). They will often be extensions of successful level 2 prototypes.

Prototypes and logical dialogue design

The level 1 dialogue simulation will give users a better way of evaluating dialogues than logical dialogue design. However, Dialogue Outlines are still useful in that they:

* provide a way of specifying dialogues if prototyping tools are not available;
* can provide a way of specifying dialogues for prototyping;
* avoid concentration on the format of the interface before the physical environment has been decided;
* can provide a way of specifying dialogues to system constructors if the prototype has to be thrown away.

SUMMARY

* The design of the on-line processing is critical to the system's success.
* Dialogue design can start in SSADM when the functions have been divided into batch and on-line catalogues in stage 2.
* SSADM uses two techniques to help specify the on-line dialogues: logical dialogue design and prototyping.
* Logical dialogue design specifies the data content and the actions performed by the user and by the computer required for the on-line processing of an event.
* Logical dialogue design is used as a communication medium between analysts and users and between analysts and constructors.
* A notation is used, based on flowcharting, which is easily understood.

- Logical Dialogue Outlines are developed in step 250 after the Entity Life Histories, in stage 5 the event level outlines are linked together by Logical Dialogue Controls.
- A suggested approach to developing Logical Dialogue Outlines involves: identifying on-line events, identifying data items, building the outline, and then validation.
- Prototyping has become an important technique in systems development.
- Variations of SSADM have been developed that incorporate prototyping, these are known as levels 1, 2, and 3 and represent: screen-based dialogue simulation (level 1), a throw away working prototype (level 2), and a working prototype that will be developed into the production system (level 3).
- Prototyping and logical dialogue design can be used separately or together within SSADM to specify the user interface.

EXERCISE

Scapegoat Systems
Create an event level Logical Dialogue Outline for the event, New Employee, described below.

When a new employee joins Scapegoat, his or her details are entered on the system.

The personnel manager enters the new employee's name, grade, and the department to which he or she will belong.The system accepts the name and allocates a staff number which is next in sequence.

The personnel manager will then indicate the person in the same department who will be the manager of the new employee.

The personal details such as address and date of birth are entered. After this a variable number of qualifications are entered (this can be none).

At the end the system confirms successful completion of the dialogue.

4.9   Reviewing the Required System Specification (step 260)

Purpose

This review is a very important one. The required system has been analysed in a great deal of detail and the users must now confirm that this is the system they want before the analysts go away and decide how best to implement it.

Stage 2: products to be reviewed

The list of products to review at the end of stage 2 is very long and it is difficult to review all items described here within one meeting. It is recommended that products such as the Data Flow Diagrams and the Logical Data Structure are reviewed at the end of the steps in which they are completed. As both of these diagrams may undergo amendment during the construction of Entity Life Histories, they must still form part of the documentation of this review. However, only the amendments made as a result of Entity Life Histories will need to be examined by the reviewers.

As the set of documentation is large and complex, the first task of the reviewers is to ensure that the documentation is complete. Some of the completeness checks that should be made are:

- All entities, except for operational masters, should have an associated Entity Life History.
- All events that appear in the Entity Life Histories should be described on the Event Catalogue.
- All bottom-level processes should have an associated Elementary Function Description.
- All entities should have an associated Entity Description.

To help the reviewers perform this check, the project team should provide a list of contents with the set of documentation given to each reviewer.

Required System Data Flow Diagrams
These diagrams must be explained fully to the user reviewers to ensure that the impacts on working practices, staffing, and so on are fully appreciated.

The members of the review team will adopt the following roles:

- The user reviewers should be sure that they understand exactly what processing the computer will be performing and what will be required of whoever operates the system. Also, the functional breakdown represented on the Data Flow Diagrams should not conflict with the potential structure of the organization.
- The technical reviewer plays a more dominant role than in the stage 1 review as it is important that this set of Data Flow Diagrams not only conforms to the rules of Data Flow Diagrams but also is consistent with the Logical Data Structure and Event Catalogue: the flows at all levels must be consistent; all data stores must be updated only by processes within the system; all events should be identifiable as triggers to the processes.

Elementary Function Descriptions
An Elementary Function Description should exist for each bottom-level process.

- The user reviewers should be happy that these brief descriptions are clear and informative about the nature of the bottom-level processes.
- The technical reviewer should check that if the events are not clearly shown on Data Flow Diagrams they are indicated on the Elementary Function Descriptions. In the case of time-based or internally recognized events, the circumstances of time-intervals should be indicated here.

Required System Logical Data Structure
The Logical Data Structure should cover the complete set of entities and relationships required by the new system. Each entity should be described, and the relationships explained.

- The user reviewers should be certain that all of the information they will require in the new system is represented here in a form that will allow them to access the data in the way that they want.
- The technical reviewer should be satisfied that all of the one-to-many relationships are true from both directions in that occurrences of master entities may be related to many occurrences of their detail entities and that occurrences of detail entities will always relate to only one occurrence of their masters. Optional relationships should be tested in the same way. It is a common error to show a relationship as optional when a master

entity optionally is related to detail entities. The optionality only applies when a detail may not be related to masters.

Entity Descriptions

Each entity on the Logical Data Structure, excluding operational masters, must be supported by an Entity Description. It is not necessary to have a complete set of data items for each entity, as these will be added at the Composite Logical Data Design step in stage 4. It is useful to have the data items that uniquely identify the entity plus other significant data items. These will help in the review of the Entity Life Histories.

Entity/Data Store Cross Reference form

This is a very important aid to cross-checking the Logical Data Structure and the Data Flow Diagrams. Without this, it is very difficult to understand exactly what is being updated on a Data Flow Diagram every time a data store is written to.

All of the entities must appear on the Entity/Data Store Cross Reference form. Although more than one entity may be related to a single data store, it is not permitted for an entity to be related to more than one data store. All main data stores should be on the Entity/Data Store Cross Reference form.

Problems/Requirements List

The Problems/Requirements List should have the majority of solutions added by this step. The solutions will have been added during the construction of the Required System Data Flow Diagrams and Logical Data Structure. Any solutions still not present at this stage will be physically dependent, so might be added as a result of technical options (stage 3). The review meeting should ensure that any changes and additions made to the list since it was signed off earlier are agreed by the user reviewers.

Function Catalogues

These define which functions are handled batch and which on-line. The users should confirm that they are happy with the way these have been assigned and with the time-frames (daily, weekly, monthly, etc.) allocated to the batch functions.

Retrievals Catalogue

This should describe each enquiry that the user will require:

- The user reviewer should ensure that none of the important enquiries have been omitted and that the enquiries specified accurately represent the needs of the users.
- The technical reviewer might ensure that the enquiries could be supported by the Logical Data Structure by asking that the presenter walk through the access paths, showing where each item of information is obtained from on the Logical Data Structure.

Entity Life Histories

It is not always useful for the users to be involved in the review of the Entity Life Histories. The diagrams are often not easy to follow, so in practice a separate technical review of the Entity Life Histories is held without user involvement. However, it is important that the information represented within the Entity Life Histories is validated. Possibly, a list of assumptions could be reviewed at the full review meeting.

Event Catalogue

The Event Catalogue is the end-product of the Entity Life Histories, and even if the Entity Life Histories themselves are not going to be reviewed here, it is important that the list of events, their descriptions, and the entities they affect are reviewed by the user and technical reviewers.

Logical Dialogue Outlines

These can only be reviewed by walking through each Logical Dialogue Outline, trying to show how the dialogue between the user and the system will work.

- The user reviewers should be sure that they understand what the Dialogues will require of the users and agree this.
- The technical reviewer might check that the updates performed within a Logical Dialogue are possible given the set of entities affected from the Event Catalogue.

I/O Descriptions

An I/O Description is eventually raised for each screen/print to be implemented within the new system. At this stage, though, the I/O Descriptions might represent the set of data items represented by particular data flows. The I/O Descriptions should be checked to ensure that the data items required by a process, event, dialogue combination are present on the I/O Description.

5. Technical options (stage 3)

5.1 Introduction

Purpose of technical options

Although it is stressed that the first five stages of SSADM are done at a logical level, it is necessary to make the major decisions about the final implementation before the detailed data and process design is done. This is because the physical implementation might very fundamentally affect the design at the logical level. For example:

- If the system is to be implemented on a large central mainframe with no constraints on size or volume, the logical design would not be constrained.
- If implemented as a network of micro-computers, the design might be constrained since the processing power and physical size would be inadequate for the more sophisticated functions of the system.
- If a distributed system is chosen, the data and functions would have to be organized in a particular way which would be reflected at the logical level.

Although the broad details of the physical implementation may have been decided upon some time previously—in a feasibility study or during the specification of requirements— the precise details are agreed here. This is only possible at this stage because the detailed knowledge about the requirements of the system will only have been gained in the specification of requirements stage.

Although it is not possible to predict how much detail must be gone into in stage 3, it is probable that a detailed cost/benefit analysis will be prepared for each option, together with a detailed system description that will enable the users to decide upon the system that will be implemented.

Inputs and outputs of stage 3

All of the information collected so far in the project will act as an input to this stage, plus research into the costs and specifications of several of the possible implementation options. For sizing purposes:

- The Logical Data Structure with related Entity Descriptions will give a rough estimate of the size of the storage required. This is only rough because until the detailed relational analysis is performed, the Entity Descriptions are likely to be incomplete. The volumes of data collected in stage 1 will give an idea of the capacity required.
- The Event Catalogue and Retrievals Catalogue will give a rough idea of the processing load. Again, the logical design might add to the events and retrievals required, so it will not be possible to obtain an accurate estimate.

Within the boundaries and constraints already laid down in stages 1 and 2, a number of possible implementation options may be constructed.

The major inputs, then, are the documents and diagrams produced in stage 2 together with technical information on possible implementation options. The chosen option is expanded to form the final Required System Specification. This is reviewed by the users to ensure that it is an accurate reflection of the system required. In addition, Performance Objectives are set within this stage. This means that the users will be required to state the maximum time for completion that is acceptable for each of the major transactions or to set system-wide targets such as 3-second response times for all on-line transactions. The outputs, then, are the detailed Required System Specification and a statement of Performance Objectives. Figure 5.1 shows the structure of stage 3.

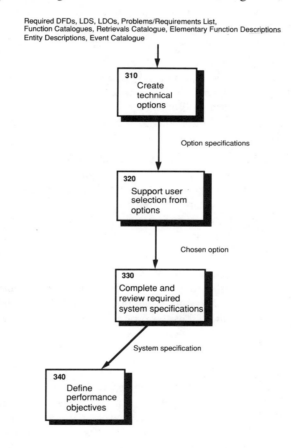

Fig. 5.1

5.2   Create technical options (step 310)

Introduction

Several options are prepared for presentation to the users who are responsible for choosing one of the options, or a solution made up of elements from several of the proffered options. This selection will be substantially supported by the project team who are able to provide technical details or an idea of the impact of certain changes to the options previously prepared.

Identification of constraints

The first step is to identify the base constraints any of the options must satisfy. The types of constraint to be considered are as follows.

Costs If there is a fixed budget for the project, it is important that none of the options exceeds this budget. Obviously, exact costs will not be calculable at this stage, but the work done in stages 1 and 2 will provide a good basis for sizing the system, and thus the type of equipment required. If some of the functions are very costly in relation to their importance in the system, it might be useful to place a cost on certain functions within the system. In this way, the users will have adequate information at their disposal to decide on what should be omitted if the costs are too great.

Timescales If there is a mandatory requirement for a quick solution, the procurement of new equipment might take too long, so implementation on existing equipment should be investigated. The use of fourth-generation software might reduce timescales considerably, but their use might constrain the sophistication of the final solution. These factors should be carefully weighed up as decisions made on the basis of timescales might limit the final solution more than is at first obvious. Unless there is a legal requirement for implementation by a certain date, the options might explore the lengthening of timescales.

Physical constraints Although not at first obvious, the physical constraints may influence the final solution significantly. For example, in a situation where a new computer is being procured, special air-conditioning might not be available, in which case a computer capable of operation in a normal office environment might be the only option available. Or if a computer is being chosen to fit on board a ship or in a court-room, the environmental aspects and the physical size of the the space available should be clearly understood.

Distribution of facilities A feasibility study or the Business System Options might already have set a constraint that the processing and/or the data should reside on more than one computer for organizational or business reasons. It is important to be clear exactly which facilities are required where, and the reasons for wanting to distribute the system. The formulation of options here might consider distributing some of the processing where local facilities are required that do not use the main data files. The requirements for communications hardware and software must be taken into account when formulating this type of option: often, this might be the constraining factor.

Facilities to be available It is worth establishing at the beginning of this stage what are the minimum facilities that must be made available, which are desirable, and which are optimal. In this way, the costs can be calculated for the minimum, optimum, and bells-and-whistles options. In the same way, certain functions of the system will be the backbone of the system, and others might be performed almost as well manually.

Service levels The service levels will have been discussed at a general level in the formulation of the Problems/Requirements List early in stage 2. The detail will now be present to be able to specify these things for individual functions, terminals, or systems where there is a distribution of processing.

Creation of the options

As a general rule, no more than three options should be prepared at this stage. If a feasibility study has preceded the analysis, the broad decisions on the size and type of solution will already have been made. Even if there is no feasibility study report to refer to, the project team should have a fairly clear view by now of what will be sensible and what will not. Something would be wrong if the options here contrasted implementation on microcomputers with that on a mainframe. Generally, the options will be different on the basis of cost, facilities available, or distribution of processing or data. As a comparison, the cost of retaining the current system (computer or manual) should be examined to be sure that the implementation of a new system can be justified. One of the unexpected side-effects of performing an SSADM analysis on a manual system has been a re-examination of the current clerical procedures, leading to the implementation of a new manual system in preference to a computer system.

The formulation of the options requires a fairly extensive knowledge of what is available.

- If a system is to be implemented on an existing computer, the hardware and software environment should be fully understood. This would be achieved by at least one of the members of the project team having spent some time as an operator or programmer within the organization.
- If a completely new system is to be purchased, a survey of what is available on the market should be conducted. This might involve obtaining literature from a number of vendors together with costs to gain a general idea of the cost of certain facilities. If a fourth-generation language or application generator is one of the options, a questionnaire could be sent to prospective vendors to find out the facilities available on each one. Demonstrations of these products should be attended also, as the facilities and man–machine interfaces vary widely over the range of development software available.

The project team may wish to present many different facilities to the users for their consideration. Obviously, it is not worth creating a new option for each facility that is to be considered, so several facilities should be included in each option. Where several facilities are mutually exclusive, a direct comparison might support the actual option presentation.

Each Option should contain the following elements:

Technical description

Even if the minute details are not worked out at this stage, it is important to sketch out the configuration of the required system in terms of central processing units, local microcomputers, and terminals. The options to do with communications between these components should also be looked at carefully. For example, what appears to be a straightforward configuration of a main central processor and several dumb terminals might require extensive work to be done in laying fibre-optic cables, in which case a replicated system on minicomputers might be preferable. Even if the system is to be implemented on an existing computer system, a technical description is still very important. Some enhancements to the existing computer or more sophisticated software controls might be required on the implementation of the new system.

Another aspect of the technical description is the software which falls into the following categories:

- system or operating system software;
- file handler or database management system software;
- development software;
- application packages;
- programs to be developed.

Description of functionality if this varies between options
This is done to demonstrate how the elements in the specification produced at the end of stage 2 are satisfied in the overall system design.

Costs and benefits
A detailed description of what constitutes a full cost/benefit analysis will not be given here. It is a well-known technique of conventional systems analysis. Obviously, the users will want to know the order of cost of each option and the benefits associated with each option. An estimation of the number of years it would take to recoup costs based on the savings expected should be drawn up. If the benefits are mostly intangible (e.g. the system will improve the quality of life for the users), then this sort of analysis might not be possible.

Impacts on the organization
The assessment of impact on the organization and environment of the users should include the following aspects.

Staffing levels   The new system may require a reorganization in the personnel. The computer may replace the function of certain staff, or increase their efficiency to the extent that staff numbers might be reduced. A less obvious source of possible discontent is where the same numbers of staff will be needed, but the distribution amongst the grades may change, or new posts specific to the operation of the computer may need to be introduced. All of these factors must be presented to the users at this stage to avoid possible industrial relations problems later.

Operating procedures   If the current system is implemented on a computer, the changes in the actual job of the users may not be very obvious. If a complete redesign of a system is not absolutely required, it is usually best to make the new system look as familiar as possible to the users to minimize the relearning required. New facilities might be introduced or existing batch input may be made on-line. In systems where a process or physical stock movement is being monitored, it is very important to have the commitment of all levels of users—not just management—at this stage to enforce the use of the system to ensure consistency and accuracy of information in the system. Without this commitment of the users, the system would be of no use to anybody.

Special training required The training required might consist simply of spending a short time with the implementer to learn the menus and commands, or if a user-tailored enquiry facility is required, the learning of a query language at the vendor's training establishment might be required. In complicated systems, the learning curve of the users may affect their productivity during the cut-over from the current system to the new system. Although the final hardware and software to be used may not have been decided at this point, the different options may show great differences in this area, so an estimate of the impacts here should be made.

Type of cut-over from current to required system If there is to be parallel running, the work of the organization might suffer for a time after the introduction of the new system while the users do their job twice—once on the current system and once on the new system. If the new system is to be implemented in phases, then this must also be made clear and the impact upon the work assessed

Enhancements/reductions in effectiveness The impact on the actual work achievable by the users should be assessed. For example, a system might be introduced to monitor a process which has not previously been monitored. In this case, it is likely that the people doing the job will become less effective because of the additional time spent inputting information to the system. The benefits to management might outweigh these considerations, but it is important for all of these things to be considered.

Timescales and resources required for implementation
A rough plan up to the 'go live' date should be drawn up for each option as a basis for estimating development costs and to ensure that the timescales are sufficiently short for the users. These timescales will include:
- completion of logical and physical design;
- procurement of hardware and/or software;
- development of programs or installation of bespoke/package software;
- testing;
- preparation of manuals and user training;
- installation of live system.

Obviously, timescales will be greatly affected if:
- equipment must be procured through an open tender procedure;
- application packages are installed on existing equipment;
- fourth-generation software is used for the development;
- the entire system is developed by an external contractor as a turnkey system.

Although more costly, some of the quicker options may be cheaper based upon cost of development time.

System sizing
The size of a system has two components: stored data and processing load.
    The size of the data can be estimated by the following procedures.
1. Add together the lengths of each data item on the Entity Descriptions.
2. Add an extra 15–25 per cent to this figure for each entity for the extra data items that might be added by relational data analysis.

3. Multiply the length of each entity by the number of times the entity occurs.

4. Add together the total size of each entity.

5. Add an extra percentage to this figure to allow for overheads and growth (the usual rule of thumb is to double).

This figure is an estimate of the storage requirements. If there is a requirement to hold historical data on the same database as the current data, then the length of time that data needs to be kept must be considered .

The size of the processing load is most relevant when expressed as a peak and average figure. It is the peak loading that will determine the size of processor required. This is determined in the following way.

1. Ensure that the volumetric information is present on the Event Catalogue.

2. Assess the maximum and average volumes of all the entries on the Retrievals Catalogue.

3. If the peak volumes occur only at certain times of the day or on certain days of the month, etc., draw loading graphs, showing where the peaks occur. (An example of this is where claims or orders are received by post: a peak of the entry of the details of the claims or orders will occur after first and second post.)

4. From the above, estimate the maximum number of events and enquiries to be handled in the worst case.

5. For the most significant events and enquiries, i.e. the most numerous at the peak:

    (a) estimate the number of I/Os (roughly, how many entity occurrences are accessed);

    (b) from an average read/write time, estimate the time for all the I/Os;

    (c) for complex dialogues, use the service time on the Logical Dialogue Outlines to estimate the time taken for each event or enquiry additional to the I/O time. (Service times are sometimes added to the Logical Dialogue Outline to show the time taken by the user to perform manual tasks associated with the dialogue.)

These figures will give guidance on the processing load at the peak of operations. From these, a rough estimate of the computer power required can be obtained.

5.3   Support user selection from options (step 320)

The selection of an option would normally be done by means of a meeting similar to a quality assurance review meeting. The team of users responsible for the selection would be given a brief description of each option several days before the meeting. The project team would explain each option, comparing and contrasting each one and leading a detailed discussion about the system. The selection itself may be resolved at this meeting or at a subsequent meeting if other parties are consulted about the possible solution. The option chosen need not be just one of the options proffered, but might be a composite of features from each of the options.

5.4   Complete and review Required System Specification (step 330)

A full system specification is produced at the end of this stage to be carried forward into the detailed design stages of SSADM. Some projects may need to produce this specification as a formal report required for internal justification, as part of internal project standards, or as the basis for commercial tenders for the hardware, software, or further development.

This specification will contain:

- Stage 2 documentation which may have been modified as a result of the option chosen
- An extended description of the option chosen, following the elements described on pages 167–70, 'Creation of the options'.

Modification of the stage 2 documentation

The products from stage 2 must be updated to show any changes at the logical level that have occurred as a result of the technical option selection. If there have been changes in the system boundary, this will affect the Data Flow Diagrams, the Logical Data Structure, and the Event Catalogue. Wherever a change is made, it is important that all related documentation is updated to remain consistent. If I/O Descriptions have not been previously completed, it is important that these are completed here as an input to relational data analysis. It is here that the Problems/Requirements List is completed. All of the problems and new requirements should have been accounted for, and the solution to each may be reviewed to ensure that nothing has been missed out.

Required System Specification

The modified SSADM end-products discussed above will form part of this specification. The remainder will follow the format of the option specification and will contain the following elements.

Technical description   This will usually be a more detailed description, which includes the system sizing, than that used in the options. Modifications to the option description may be required if a 'hybrid' option was chosen.

Functional description   The products reviewed at the end of stage 2, modified as described above by the selected option.

Organizational impact   Any changes to the user organization structure and staffing levels. Impact on the users of the new operating procedures and a description of the training required for them. A clear statement of the business objectives to be gained by the new system is made for general communication and later verification against actual achievement.

Cost/benefit analysis   This may need to be modified to reflect the option selected.

Development plan   This will be an extension of the timescales and resources estimated for the chosen option. Detailed plans will be drawn up for the design phase of SSADM, and for the procurement of hardware and software (if required). Outline plans will drawn up for the construction and implementation phases of the project.

5.5   Define Performance Objectives (step 340)

Introduction

This step defines the Performance Objectives that the implemented system must achieve, these will become the targets for the physical design to reach in stage 6. This is the earliest point in the project that these objectives can be produced since they need to be based on the technical environment selected. As these objectives are not used until the physical design their development is sometimes postponed until the end of stage 5. They can then be based on more detailed information about the data structure and access requirements of the processing. However, many projects need to issue a document (sometimes called an operational requirement) to potential suppliers on which they will base their tenders for hardware and software. This document must contain details of the Performance Objectives of the system. It is for this reason that this is included in the technical options stage.

Normally Performance Objectives are produced by the development team in consultation with the users. If the system is to be implemented on existing hardware and software, then the installation's operations management will need to be involved.

The approach to setting these objectives will depend to a large extent on whether the new system is to go onto existing hardware or whether new hardware and software are to be purchased. In the first case the objectives will be constrained by the extent to which the current system utilizes that capacity. In the second case the objectives will be constrained by the technical environment selected and the budgets agreed for the project.

Data storage criteria

The aspects that will need consideration are: the amount of main memory space available, the space available on the various kinds of backing storage, and any limits imposed by the system software. The amount of main memory space may be limited by the hardware employed, by the project budgets, and by the constraints of other systems using the same hardware. Limits to buffer space may constrain other space-related issues such as the page or block size as well as having an effect on throughput. The amount of on-line disk space available may also be constrained by the hardware, project budgets, and by the requirements of other systems. The database management system and the operating system may also impose limits on the page or block size, the total addressable space, the number of data items per record, and the way that variable length items are handled.

At present, space is rarely a serious problem as costs both of disk storage and main memory have dropped considerably in the last ten years. Problems are most likely to occur when the limits of existing hardware are reached by the installation. However, even when new hardware and software are to be purchased it is still a good idea to set performance criteria for data storage. This then gives the physical designers a target to ensure efficient usage is made of resources. This objective can be based on the system sizing developed in the specification of the chosen option.

Function timing criteria

Often projects set general objectives that are applied across the whole system and are based on no real analysis of what is achievable. Typical is the statement that 'All on-line response times will be less than 5 seconds'. These sort of ill-thought objectives lead to great frustration when it becomes obvious (either through performance estimation in stage 6 or, much worse, when the system goes live) that they are impossible with the configuration selected.

Thus it is important to base the objectives on the system sizing developed in the specification of the chosen option and to analyse the requirements for each critical function separately. The critical functions usually belong in these categories:

- high volume on-line transactions;
- major daily batch runs;
- programs run at peak times;
- on-line transactions requiring a fast response or real-time transactions.

The selection of critical functions is discussed in more detail in on pages 246–7, where the selection of critical programs for initial specification is discussed.

Batch function timing objectives are often predetermined by the business objectives of the system and the turnaround time expected by users. Thus if invoices have to be sent on the first day of the month, then they have to be processed as part of the month end cycle. A proportion of the machine time allotted to that cycle will then become the objective for the batch function that produces invoices. In some cases it may be necessary to base the objective on an estimate of the resource requirements for the function.

On-line function timing objectives can be set by taking the total machine resources available and dividing it by the total number of transactions at peak times; this will give a rough figure for the resource allocation due each transaction. More accurate figures can be determined by basing the calculation on an estimate of the number of disk accesses required by each transaction. At the end of stage 5 the figure given on the Process Outlines will help with this estimate. Another approach to calculating a reasonable objective is to agree a response time with the users, say less than 10 seconds, and to subtract from it estimates of other contributing factors such as the line speeds and queueing time. This will give an estimate of the acceptable resource time for all transactions. However, whatever method is used some previous sizing must have been done to ensure that the objectives are achievable.

Information objectives

The information objectives are implied by the Logical Data Structure and later in the project by the Composite Logical Data Design. However, as a result of the tuning performed in stage 6, access to certain information might be reduced or sacrificed. For instance, if there is a space constraint one way of meeting it will be to archive data earlier. Thus the space objectives may be competing with the information objectives. It is therefore necessary to agree with the users those elements of the system's data to which access should not be reduced.

Recovery criteria

To some extent the recovery criteria will have been set by step 205, *define audit, security, and control requirements*. Figures should be agreed with the users for mean times between and maximum times to recover from various types of failure. The performance levels of any back-up systems should also be agreed with users.

Other objectives

Any other performance objectives not discussed above need to be specified. These will vary greatly between projects but may include: operating environment conditions (e.g. if the hardware is to be situated in an engineering workshop), data communications line speeds, portability of the software, database management system software characteristics such as hardware independence, reorganization timings, and the performance of interface with other systems.

SUMMARY

- Before detailed design can begin it is important to agree on the physical architecture of the new system.
- Base constraints are identified which all options must satisfy.
- The base constraints will cover such things as costs, timescales, the physical environment, distribution, and basic facilities.
- Usually three technical options are prepared and presented to the users.
- Each option will contain details of: the technical architecture, system sizing, functionality supported, costs and benefits, impacts on the organization, rough plans for construction and implementation.
- An option, or combination of aspects from several options, is selected by the senior users with the help of the development team.
- Any changes to the stage 2 documentation necessary are made.
- A Required System Specification is developed from the modified stage 2 documentation and the selected option specification; this may be incorporated into a formal report.
- Performance Objectives are set for data storage, function timing, information, and recovery.

6. Data design (stage 4)

6.1 Introduction

In stage 2 of SSADM we were concerned with defining the major objects about which data was to be held (entities) and their interrelationships. These were documented by a Logical Data Structure diagram supported by Entity Descriptions detailing the data item contents of each entity. The Logical Data Structure was produced by a top-down approach of identifying the entities and relationships. The major data items were then assigned to each entity. This model was validated by informally checking that the processing for the required system could be performed on it.

Objective of stage 4

The objective of this stage 4 is to produce a complete detailed logical data design that can be used as the basis for file or database design in stage 6 (physical design). This design must specify all the necessary data items, grouped together into data groups in the most optimal way, and specify all the relationships between these data groups and possible access paths. This data design is known as the Composite Logical Data Design.

Comparison of logical data structuring with relational data analysis

The Logical Data Structure produced in stage 2 has some major limitations which mean that, by itself, it is unsatisfactory as the final data design. These are:

- Only the major data items were identified for each entity.
- A sub-optimal, inflexible assignment of data items to entities may have occurred since formal techniques were not used.
- There may be undiscovered inherent relationships between data items which could be invaluable for some retrievals.

Logical data structuring	Relational data analysis
Top-down	Bottom-up
Based on analysis of entities and their interrelationships	Based on analysis of data items and their interrelationships
Intuitive subjective technique	Formal, rigorous, mathematically based technique
Based on and validated against the processing requirement	Based on the data contents of system input and outputs
May produce simple, inflexible structures	Produces highly flexible, complex structures
Model is represented as a diagram showing entities and relationships	Model is represented by groups of data items, with key items identified known as normalized relations

Fig. 6.1

The degree to which the Logical Data Structure, produced in stage 2, will be subject to these limitations is quite dependent on the effort expended then.

To overcome these limitations an additional approach to defining the system's data is used, relational data analysis. This is a bottom-up technique in that the relationships between data items are rigorously analysed in a way that is independent of the processing. This analysis results in groups of data items organized in a flexible, non-redundant way. These are known as normalized relations. The differences between relational data analysis and logical data structuring are summarized in Fig. 6.1.

Structure of stage 4

Stage 4 is made up of two steps: step 410, *carry out relational data analysis,* and step 420, *create detailed logical data design.*

To produce the detailed logical data design the results of relational data analysis are combined, in step 420, with the required Logical Data Structure. Thus the strengths of both techniques are used to ensure that the Composite Logical Data Design is as complete and accurate as possible. The Composite Logical Data Design is expressed as a Logical Data Structure style diagram supported by a full definition of the data contents of each data grouping. For each data grouping primary key data items and foreign key data items are identified. These key data items enable the identification of each data group occurrence and enable the accessing of other data group occurrences through common key data items. Full volumetric information is specified. The data space required for each data group occurrence, total number of occurrences for each data group, and the numbers of occurrences participating in each relationship are all recorded.

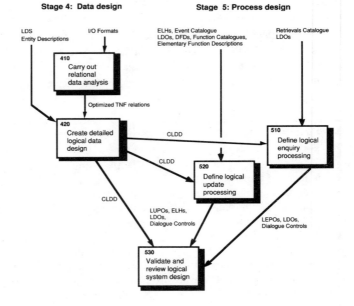

Fig 6.2

In stage 5 the processing of the new system is defined in detail. The processing for each update and enquiry is defined in terms of the operations performed on the data and in terms of the navigation of the Composite Logical Data Design necessary to obtain that data. In this way the Composite Logical Data Design is rigorously validated against the processing requirement. A formal quality assurance review is carried out in step 530, *validate and review logical systems design.*

Figure 6.2 shows the steps involved in stages 4 and 5.

6.2   Relational data analysis (step 410)

Introduction

This section is separated into two parts. The first part deals with relational data analysis in a general sense, explaining the background, underlying theory, terminology, and demonstrating the technique. The second part deals with the approach taken to relational data analysis in SSADM, illustrating this with examples from the Yorkies case study.

Background to relational data analysis

Relational data analysis is a very widely used technique, and several modern systems development methodologies use it in some form. The ideas behind it stem from theoretical work published by Codd in the early 1970s. This work led to the development of relational database theory and to the development of the relational data analysis technique for database design. The basic ideas of viewing all data in simple tables (or relations) is common to both relational database theory and to relational analysis. The database theory concentrates on the manipulation of these tables, whereas relational analysis concentrates on how data might best be organized into these tables. The approach described in this section is a simplification of some of the more abstract and theoretical ideas of relational theory (for a full treatment the reader is referred to the database books listed in the bibliography and the references they contain).

Tables, columns, and rows

The product of relational data analysis will be a set of tables in which all of the systems data can be represented. These tables are often known as relations. (The terminology is confusing; relations are nothing to do with relationships in Logical Data Structures, in fact, they are equivalent to the entities.) The mathematical origins of relational theory give rise to a number of specific mathematical terms whose meaning is not obvious to the less mathematically minded. We use the more informal equivalents but give the formal relational terms.

Figure 6.3 shows two tables from the hospital example used in Chapter 2.

Table or relation

This comprises both the actual data occurrences and the heading information at the top of the table.

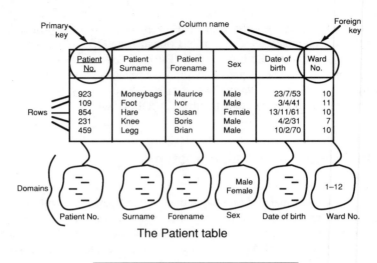

The Patient table

Ward No.	Ward Name	Type	Number of beds
3	Nightingale	Medical	8
11	Fleming	Medical	12
10	Barnard	Surgical	21

The Ward table

Fig. 6.3

Rows

The rows of the table (tuple is the formal relational term) show the various occurrences of the patient. For instance, the first row shows the various values associated with Maurice Moneybags' stay in hospital.

Two important properties of relations are concerned with rows.

1. In any table there can be no duplicate rows. It follows from this that each row must be uniquely identifiable. In the most complex case it may require a combination of all the values in the row to identify it. However, in most cases it requires one value or a combination of two or three values to identify each row. In the case of the Patient table it requires only the value of the Patient No. to identify the row. The column name (or names) that identify each row are referred to as the primary key. Thus Patient No. is the primary key of the Patient relation and Ward No. the primary key of the Ward relation. As there can be no duplicates the primary key must always exist.

2. The order in which the rows appear in the table must *not* be significant. This follows from the principle that in relations all data is represented explicitly, i.e. in terms of values rather than position. Thus if the Patient table was in order of when the patients were to be operated on, then this should be shown explicitly by adding a new column, Operation Sequence No.

Columns

Each column has a heading or column name (the formal term is attribute) and a set of values that are taken by the attribute in different rows. The column name is equivalent to the data item type and each value is equivalent to a data item occurrence in the Logical Data Structure.

Two important properties of columns in relations exist.

1. The order of columns must not be significant. We have shown, as is the convention, the primary key columns as the first column in the table, but the relation has to be represented on paper in some order. This property is less important in relational analysis than in relational database theory where the unordered nature of the relations enhances data independence.

2. The second important property is that only one value should be associated with each column/row intersection in the table. So if we were to add a column to the patient table in Fig. 6.3 of Drug Prescribed, then because a patient may be prescribed several drugs during their stay there could be several values for, say, Maurice Moneybags' Drug Prescribed, e.g. penicillin, aspirin. This breaks the rule of one value per column/row intersection. Another way of expressing this rule is that there can be no repeating groups of data item occurrences for one occurrence of the primary key. This rule is important as it means that a relation must at least be in what is called *First Normal Form*. The normal forms of data are discussed later.

Domain

This represents a pool of possible values from which the actual values appearing in the columns of the table are drawn. Thus the domain of Patient Nos. includes all of the possible Patient Nos., not just the ones currently in the hospital. The domain of Sex has just two possible values—male and female. The importance of domains is shown when we compare values from different tables. Thus if we wanted to find the name of the ward a particular patient was in we would have to compare the values of Ward No. in the Patient table with the values of Ward No. in the Ward table. This could only sensibly be done if the two Ward Nos. came from the same pool of values.

Another example of this could be if we had another table called 'Doctor' which had column names Doctor No, Doctor Surname, Doctor Forename, Doctor Date of Birth, etc. To find out if we had any doctors who might also be patients we could compare the attribute values: Patient Surname with Doctor Surname, Doctor Forename with Patient Forename, Patient Date of Birth with Doctor Date of Birth. This can be done because the columns being compared share the domains: surname, forename, date of birth.

Normalized relations

The object of relational data analysis is to organize all of the system's data items into a set of well normalized relations. Well normalized relations avoid certain undesirable properties:

- unnecessary duplication of data items in different relations (i.e. no redundant data);
- problems with modifying, inserting, and deleting data (sometimes referred to as the update anomalies).

To organize data into well normalized relations we go through several stages of normalization known as normal forms:

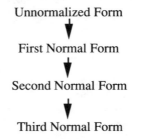

There are additional normal forms but, in practice, these are very rarely met and are not discussed here or in the SSADM reference manual (the interested reader should consult Date (1986) or other comprehensive database textbooks).

As the data are analysed to Third Normal Form the technique is often referred to as Third Normal Form (TNF) data analysis.

Example of relational data analysis

Relational data analysis can be applied to any set of data items. One could , theoretically, take all of the data items of the system, analyse them, and produce a data design for the system all in one step. In practice this is almost impossible to do (unless the system is very small) since it involves comparing every data item with every other. The practical approach taken in SSADM is to analyse separately small 'chunks' of the system data and to synthesize the results of each separate analysis to produce a set of relations for the whole system. The 'chunks' of data selected for analysis are often referred to as *ata sources*. Normally the inputs and outputs from the system are used; these could be forms, screen formats, or reports.

To illustrate the technique we will extend the hospital example described earlier. Suppose that each patient has, at the end of their bed, a Drug Card showing all drugs prescribed during their stay and the dosage required. The Drug Card is therefore our data source and two sample cards, with information typed onto them, are shown in Fig 6.4. These examples will be used to represent all of the drug card data for all patients. It is important to realize we are analysing all of the existing data and possible data associated with the data source chosen.

1. Represent the data in unnormalized form and pick a key The first step is to represent all of the data in a table. Some of the column headings (data item names) have been abbreviated to save space. As the dates given on the Drug Card are the dates on which the particular drug was prescribed we have renamed Date to the more meaningful Prescription Date. Note that the table is not strictly speaking a relation since there are several possible values for, say, patient 923's Drug Code.

Drug Card

Patient No.: 923   **Surname:** Moneybags   **Forename:** Maurice

Ward No.: 10   **Ward Name:** Barnard

Drugs prescribed:

Date	Drug Code	Drug Name	Dosage	Length of Treatment
20/5/88	CO2355P	Cortisone	2 pills 3 x day after meals	14 days
20/5/88	MO3416T	Morphine	Injection every 4 hours	5
25/5/88	MO3416T	Morphine	Injection every 8 hours	3
26/5/88	PE8694N	Penicillin	1 pill 3x day	7

For additional drugs continue on another card

Drug Card

Patient No.: 109   **Surname:** Foot   **Forename:** Ivor

Ward No.: 11   **Ward Name:** Fleming

Drugs prescribed:

Date	Drug Code	Drug Name	Dosage	Length of Treatment
15/5/88	AS473A	Aspirin	2 pills 3 x day after meals	7 days
20/5/88	VA231M	Valium	2 per day	5

For additional drugs continue on another card

Fig. 6.4

We can select any data item or combination of items to act as primary key. However, it makes the analysis rather more straightforward if a 'reasonable' key is selected. Criteria for selecting a reasonable key are:

- ideally the key should be unique (only one possible value) for the particular data source;
- use the smallest combination of items possible;
- avoid textual keys.

The Patient No. is a good choice since each Drug Card is for one and only one patient, and the other criteria are satisfied. The convention is that primary keys are underlined. Figure 6.5 shows the Drug Card data in unnormalized form.

Pat No.	Surname	Fore-name	Wd No.	Ward Name	Prescr Date	Drug Code	Drug Name	Dosage	Lgth Treat
923	Moneybags	Maurice	10	Barnard	20/5/88	CO2355P	Cortisone	2 pills 3 x day after meals	14
					20/5/88	MO3416T	Morphine	Injection every 4 hours	5
					25/5/88	MO3416T	Morphine	Injection every 8 hours	3
					26/5/88	PE8694N	Penicillin	1 pill 3 x day	7
109	Foot	Ivor	11	Fleming	15/5/88	AS473A	Aspirin	2 pills 3 x day after meals	7
					20/5/88	VA231M	Valium	2 per day	5

Fig. 6.5

2. Represent the data in First Normal Form by removing any repeating groups of data items to separate relations. Pick keys for any relations identified A repeating group is defined as any data item or group of data items that may occur with multiple values for a single value of the primary key data item.

Thus in the table shown in Fig. 6.5 there are several values of Drug Code, Drug Name, Prescription Date, Dosage, and Treatment Length for one value of the Patient No. These items form a repeating group and are removed to a separate relation.

The new relation has the column headings Patient No., Drug Code, Drug Name, Prescription Date, Dosage, and Treatment Length. The Patient No. is required to make each row unique across the whole of the system data; it is quite likely that two patients could be given exactly the same prescription on the same day.

We now have to decide on the primary key of the new relation. This will always be a combination of the key selected in step 1 and one or more additional data items necessary to identify a particular row. Patient No. is therefore part of the key and further analysis shows that it is necessary to have both Drug Code and Prescription Date included in the key. (A patient may be prescribed several drugs on the same day or may be prescribed same drug on separate occasions.) When several data items are required for the key of a relation this is known as a *compound key*. The new relation shown in Fig. 6.5 has a compound key of Patient No., Drug Code, and Prescription Date.

Pat No.	Presc Date	Drug Code	Drug Name	Dosage	Lgth Treat
923	20/5/88	CO2355P	Cortisone	2 pills 3 x day after meals	14
923	20/5/88	MO3416T	Morphine	Injection every 4 hours	5
923	25/5/88	MO3416T	Morphine	Injection every 8 hours	3
923	26/5/88	PE8694N	Penicillin	1 pill 3 x day	7
109	15/5/88	AS473A	Aspirin	2 pills 3 x day after meals	7
109	20/5/88	VA231M	Valium	2 per day	5

Fig. 6.6

With the repeating group removed to a separate relation, we now consider the data items left behind. These are the data items that do not repeat for a single value of the key selected in step 1. Each row is therefore uniquely identified by the value of the key previously selected.

Thus Patient No., Surname, Forename, Ward No., and Ward Name remain as a relation with a key of Patient No. This is shown in Fig 6.7. The data is now represented by two tables in First Normal Form.

Pat No.	Surname	Fore-name	Wd No.	Ward Name
923	Moneybags	Maurice	10	Barnard
109	Foot	Ivor	11	Fleming

Fig. 6.7

3. Represent the data in Second Normal Form by removing any data items that only depend upon part of the key to separate relations This only affects relations that have compound keys. We have to decide whether any data items in a compound key relation are dependent on only part of that compound key.

This concept of dependency, often referred to as *functional dependency*, is very important in relational data analysis.

For any two data items A and B, A is dependent on B if and only if:

> For a given value of B there is associated with it
> precisely one value of A at any one time.

Thus the data item Patient Surname is dependent on the data item Patient No. since for a given value of Patient No., say 923, there is associated with it precisely one value of Patient Surname, in this case Moneybags.

Another way of describing this is to say that:

> Data item B *determines* data item A
>
> Patient No. *determines* Patient Surname

Notice that the opposite is false:

Patient Surname *does not determine* Patient No.

For a given value of Patient Surname, say Moneybags, there may be several associated values of Patient No., as there may be several patients called Moneybags in the hospital at the same time.

Dependency can also occur with groups of data items. In the table shown in Fig 6.7 the combination of data items—Patient No., Prescription Date, and Drug Code—determines each of the data items—Dosage and Treatment Length. However, the data item Drug Name is not determined by the whole key but only by the Drug Code.

Thus the relation is subject to step 2 and the data item, Drug Name, which is only dependent on part of the key, is removed from the relation. Drug Code and Drug Name form a new relation with Drug Code as the key. The modified relation and the new relation are shown in Fig. 6.8.

Pat No.	Presc Date	Drug Code	Dosage	Lgth Treat
923	20/5/88	CO2355P	2 pills 3 x day after meals	14
923	20/5/88	MO3416T	Injection every 4 hours	5
923	25/5/88	MO3416T	Injection every 8 hours	3
923	26/5/88	PE8694N	1 pill 3 x day	7
109	15/5/88	AS473A	2 pills 3 x day after meals	7
109	20/5/88	VA231M	2 per day	5

Drug Code	Drug Name
CO2355P	Cortisone
MO3416T	Morphine
PE8694N	Penicillin
AS473A	Aspirin
VA231M	Valium

Fig. 6.8

The systematic way of deciding whether there are any part-key dependencies is to go through each compound key relation and ask of each data item:

*Does it depend on the **whole** key?*

If the answer is *No* then which data item(s) it depends upon must be decided. An additional relation is created with the determining item(s) as the key and the dependent data item(s) as the data. The dependent items are removed from the original relation.

In the example Drug Name is only dependent on Drug Code, so an additional relation is created with Drug Code as key and Drug Name as data. Drug Name is removed from the original relation.

Note that any relations that have single keys are not affected by step 3. They are already in Second Normal Form.

4. Represent the data in Third Normal Form by removing any data items not directly dependent on the key to separate relations This step is similar to the previous one in that we are looking at dependency between data items. The difference is that here we are looking for data items that might be dependent on other data items instead of looking for non-key data items that might be dependent on only part of the key.

Therefore for each data item we should ask the questions:

> *Is the data item directly dependent on the key data*
> *item(s) of the relation it is in?*

and

> *Is it directly dependent on any other data item(s)*
> *in the system?*

If the answer to the first question is *No* then the answer to the second must be *Yes*. These two questions act as a cross-check on each other in trying to find any undetected inter-data dependencies. These are always difficult to find since to ask the second question systematically we would have to compare each data item in the system with every other item or combination of items. This systematic approach is obviously impossible in practice so a more intuitive approach has to be taken. The intuitive approach relies on the analyst's skill and knowledge of the system in being able to spot possible inter-data dependencies and then to investigate them formally. In most cases the inter-data dependencies will be obvious and will occur within the relations being analysed together—it is those rare cases where a dependency occurs between items analysed on separate data sources that are hard to find.

(The technically minded reader should note that these questions ensure that the relations are in Boyce–Codd Normal Form, sometimes referred to as Strong Third Normal Form. We are following the conventions described by the SSADM reference manual; which makes no distinction between Third Normal Form and Boyce–Codd Normal Form, and describes the latter as Third Normal Form.)

If an inter-data dependency is detected, say between data items A and B, then we must decide whether item A determines item B or vice versa. The following questions should be asked:

> *Given a value for data item A is there just one*
> *possible value for data item B?*

and

> *Given a value for data item B is there just one*
> *possible value for data item A?*

If the first question is answered *Yes* and the second *No* then item A determines item B. Data item A is then key to data item B. Item B is removed from the relation it was previously in to a new separate relation whose key is item A. Item A is left in the previous relation.

If the first question is answered *No* and the second *Yes* then item B determines item A.

If both questions are answered *No* then there is no inter-data dependency.

If both questions are answered *Yes* then the two items determine each other and either could act as key to the other. These are sometimes known as *logical synonyms*. Normally the more appropriate item would be selected as the key and it would be treated in a similar way to the first case above. Occasionally it happens that the two items are also logical synonyms of the primary key, in this case no significant redundancy is involved and they can both be left in the Second Normal Form relation.

In this example examining our three Second Normal Form relations indicates a possible inter-data dependency in the relation with the key Patient No. This relation is repeated in Fig 6.9.

Pat No.	Surname	Fore-name	Wd No.	Ward Name
923	Moneybags	Maurice	10	Barnard
109	Foot	Ivor	11	Fleming

Fig. 6.9

Each non-key data item is dependent on the primary key of Patient No. in that there is precisely one value of each item associated with a given Patient No. However, there appears to be an inter-data dependency between Ward No. and Ward Name. In other words Ward Name does not seem to depend directly on Patient No. Asking the questions:

Given a value for Ward No. is there just one
possible value for Ward Name?

Yes, each ward can only have one name.

Given a value for Ward Name is there just one
possible value for Ward No.?

No, in this hospital it is possible for several wards to share a name.

This means that Ward No. determines Ward Name and therefore forms a new relation with Ward No. as the primary key. Ward No. is left in the Patient relation as its value is determined by the Patient No. Ward No. is acting as a *foreign key* in the Patient relation.

Pat No.	Surname	Fore-name	Wd No.
923	Moneybags	Maurice	10
109	Foot	Ivor	11

Wd No.	Ward Name
10	Barnard
11	Fleming

Pat No.	Presc Date	Drug Code	Dosage	Lgth Treat
923	20/5/88	CO2355P	2 pills 3 x day after meals	14
923	20/5/88	MO3416T	Injection every 4 hours	5
923	25/5/88	MO3416T	Injection every 8 hours	3
923	26/5/88	PE8694N	1 pill 3 x day	7
109	15/5/88	AS473A	2 pills 3 x day after meals	7
109	20/5/88	VA231M	2 per day	5

Drug Code	Drug Name
CO2355P	Cortisone
MO3416T	Morphine
PE8694N	Penicillin
AS473A	Aspirin
VA231M	Valium

Fig. 6.10

The two new relations arising from the Second Normal Form relation are shown in Fig. 6.10 with the other relations. These Prescription and Drug relations, developed at Second Normal Form, were already in Third Normal Form.

If we had answered to the second question: *'Yes, for a given value of Ward No. there is only one possible value of Ward Name'*, then we could select either Ward No. or Name as the key of the new relation. Ward No. is preferable as it is a non-textual key.

Note that we should also examine the key items and ask for a compound key whether any parts of the key are directly dependent on other parts of the key? If any parts of the key are dependent on other parts then the dependent items should be relegated to non-key items with the determining items remaining as keys. In the case of our compound key in the hospital example there is no inter-data dependency within the key. However, inter-data dependencies within keys are quite common, particularly in key-only relations, and we shall meet some in the Yorkies case study.

The process of normalization is now complete and to ensure that the data is well normalized two tests are applied.

5. Apply the TNF tests to check that the relations are well normalized

TNF Test 1:   *Given a value for the key(s) of a TNF relation, is there just one possible value for each data item in that relation?*

Applying this test to each of the relations defined above we ask:

> *For a given value of Patient No. is there just one possible value of: Patient Surname? Patient Forename? Ward No.?*

> *For a given combination of values for Patient No., Date of Prescription, and Drug Code is there just one possible value of: Dosage? Length of Treatment?*

Similar questions are asked for the relations identified by Ward No. and Drug Code.

To all these questions the answer is *Yes*, indicating that the relations are in First Normal Form and that each data item is dependent on the key of the relation (but not necessarily wholly and directly dependent on the key).

TNF Test 2:   *Is each data item in a TNF relation directly and wholly dependent on the key(s) of that relation?*

Applying this test to each of the relations defined above we ask:

> *Is Patient Surname/Patient Forename/Ward No. directly and wholly dependent on Patient No.?*

> *Is Dosage (Length of Treatment) directly and wholly dependent on Patient No., Date of Prescription, and Drug Code?*

Similar questions are asked for the relations identified by Ward No. and Drug Code.

To all these questions the answer is *Yes*, indicating that the relations are in Second and Third Normal Form.

To illustrate further the use of these tests we will modify the hospital example slightly. Suppose each patient belongs to a Patient Weight Type which can take values 'Very Underweight', 'Underweight', 'Normal', 'Overweight', and 'Obese'. The amount of each drug given to a patient will depend on which Weight Type they are.

Applying the two tests to the Dosage data item:

> *For a given combination of values for Patient No.,*
> *Date of Prescription, and Drug Code is there just*
> *one possible value of Dosage?*

Yes, the relation is in First Normal Form and Dosage is dependent on the key items.

> *Is Dosage directly and wholly dependent on*
> *Patient No., Date of Prescription, and Drug*
> *Code?*

No, Dosage is directly dependent on a combination of Drug Code and Patient Weight Type. Thus the relation is not in Third Normal Form and the Dosage data item should be removed from the relation to a new relation with a compound key of Drug Code and Patient Weight Type.

This kind of redundancy is quite common and is known as *transient dependency* since the dependency has been transferred from one item to another. In this case the Patient Weight Type is dependent on the Patient No. Test 2 detects the transient dependencies.

Test 2 can be rephrased to the relational oath for each data item in a relation:

> *'I swear to be dependent on the key, the whole*
> *key, and nothing but the key so help me Codd'*

SSADM notation

We have illustrated the process of normalization using sample tables of data. Although this technique makes the process easy to visualize it is rather cumbersome when analysing large tables and many data sources. A simpler notation is to use only the column headings (data item names) from the tables. These are listed down the page with primary keys underlined and shown at the top of each relation.

System: Hospital		Date / /	Author	
Source ID No.:		Name of source: Drug Card		
UNF	**1NF**	**2NF**	**3NF**	
Patient No.	Patient No.	Patient No.	Patient No.	
Patient Surname	Patient Surname	Patient Surname	Patient Surname	
Patient Forename	Patient Forename	Patient Forename	Patient Forename	
Ward No.	Ward No.	Ward No.	* Ward No.	
Ward Name	Ward Name	Ward Name		
Prescription Date			Ward No.	
Drug Code	Patient No.	Patient No.	Ward Name	
Drug Name	Prescription Date	Prescription Date		
Dosage	Drug Code	Drug Code	Patient No.	
Length of Treatment	Drug Name	Dosage	Prescription Date	
	Dosage	Length of Treatment	Drug Code	
	Length of Treatment		Dosage	
		Drug Code	Length of Treatment	
		Drug Name		
			Drug Code	
			Drug Name	

Fig. 6.11

6. Optimize relations obtained from all data sources Relational data analysis is performed on a number of data sources and will yield from each source a number of normalized relations. The process of combining those relations, to form the full set of normalized relations, is known as *optimization*. All relations that have the same primary key are merged together and the new merged relation given a name. Figure 6.12 shows the four relations obtained by normalization of the Drug Card and the two relations shown previously in Fig. 6.3 being merged.

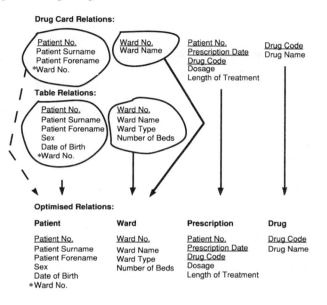

Fig. 6.12

After optimization has been performed, the TNF tests described previously should be performed to ensure that the merged relations are still in Third Normal Form.

Relational data analysis in the Yorkies case study

The previous sections illustrated the use of relational data analysis with a simplified example. We now demonstrate its use in a more typical SSADM environment using the case study.

Data sources for relational data analysis

The success of the data data analysis step in SSADM is greatly dependent on the sources selected for analysis. If insufficient or unrepresentative sources are selected then not all data items will be discovered and potentially useful data interrelationships may be missed. If too many sources are selected then the task can be enormous and exhausting.

Sources in stage 1 Relational data analysis can be used in stage 1 of SSADM when it is difficult to identify entities and produce the Logical Data Structure. The sources for analysis will then be the documents of the existing system such as the input forms, output forms, and reports. The files of the existing system are also a very useful source in that they should contain much of the data used by the existing system.

The objective of the analysis in stage 1 is to identify the major entities and their relationships—an exhaustive analysis to discover all data items is undesirable at this stage. The selection of sources should then be restricted to a few major inputs, outputs, and files. After relational data analysis has been performed on these, the optimized relations are converted into a Logical Data Structure diagram—this technique is described in the next chapter.

Sources in stage 4 In stage 4 the objective of relational data analysis is rigorously to define all the data items of the system and their interrelationships. The sources for analysis are taken from the Required System Specification developed in stage 2. It is important that all relevant sources are identified and analysed to ensure a complete and rigorous data design.

The major data sources for relational data analysis will be the I/O Descriptions produced during step 230. These detail the data contents of each major input to and output from the system. Sometimes formats or actual screens will have been produced as part of a prototyping exercise in stage 2; these will also provide valuable data sources. In general, the more structured the definitions of system inputs and outputs are, the easier they will be to analyse. This is because the underlying semantics of the data are often apparent from the way that the data is structured on the inputs and outputs.

It is often suggested that a good source for analysis are the required system Entity Descriptions, which detail the data content of each entity. However, in most projects these will already have been informally normalized. As the Entity Descriptions will be compared with the normalized relations in step 420 to produce the Composite Logical Data Design there seems little point in prejudging this activity in step 410.

The I/O Descriptions selected for relational data analysis from the Yorkies case study are listed below:

Booking Request and Confirmation
Daily Departure and Return List
Local Office Vehicle Report
Vehicle Location Report
Vehicle Relocation Booking Screen
Invoice
Reminder
Payment Input Screen
New Vehicle/Vehicle Amendment Screen
Vehicle Out of/Into Service Screen
New Driver/Amendment Screen

Normalization of Booking Request
We will systematically apply the rules of normalization to the system input Booking Request, whose I/O Description is repeated in Fig. 6.13. The result of this analysis is shown in Fig. 6.14, which shows each of the stages of normalization.

I/O Description				
System : Yorkies	Sub-system :		Author :	Date :
Name: Booking Request		Ref No.: a-1	Type: Form / screen	
Description: Customers send in a Booking Request Form or may give the same information over the telephone. In either case the information is entered into the system on-line to create the booking.				
Data items	**Format**	**Length**	**Comments**	
Customer No. Customer name Vehicle Category Code Required Booking start date Required Booking finish date Driver requirement (Y/N) Office No. (Start) Office name (Start) Office No. (Finish) Office name (Finish)			The next group of data items may be repeated if there are several Bookings { only inserted if { one-way hire	

Fig. 6.13

1. Represent the data in unnormalized form and pick a key Following the SSADM notation described previously, we list the data items in an unnormalized form (UNF). Naming of data items can cause quite a problem. Ideally data items should be named once, the name and the associated data definitions being held in the Data Catalogue. The same name should then be used whenever the same item appears on an Entity Description or an I/O Description. This consistency is easy to manage on small projects when computer support tools are used, but with large teams communication can be a problem, particularly if no multi-user computer support facilities are available. If actual screen, report, or form designs are used as data sources then the names of fields on the designs may not correspond to the names used on other SSADM documentation—care must be taken to ensure that consistent names are used for analysis. Whatever support tools are employed some sort of naming convention should be used, giving guidelines on standard abbreviations and on the position of frequently used words (e.g. date, name, description) within data item names. Some abbreviations have been used in the example below e.g.*Vehicle* becoming *Veh, Required* becoming *Req,* and *Category* becoming *Cat.*

Notice that the data item, Booking No., has been added to the list of UNF items. This kind of addition is very common when analysing system inputs. We will only hold confirmed bookings on the system so there is little point in analysing an enquiry prior to update when by adding the Booking No. we can analyse the update. This is equivalent to analysing the Booking Confirmation sent to the customer.

As each customer will presumably have only one Booking Request with Yorkies at any time, we have selected Customer No. as the key. This has been underlined, thus following the SSADM convention.

2 Represent the data in First Normal Form by removing any repeating groups of data items to separate relations. Pick keys for any relations identified There is a repeating relation for each booking the customer has made. This is removed to a new relation. Part of the key must be the key defined in step 1, Customer No., and the other part(s) must identify each occurrence of the repeating group. The only item that can do this is Booking No, so the compound key of the new relation is Customer No. with Booking No. The only item that does not repeat is Customer Name so this is left in the relation identified by Customer No.

3. Represent the data in Second Normal Form by removing any data items that only depend upon part of the key to separate relations Second Normal Form is only applicable to compound key relations, so we ask:

> *Do any of the items in the relation identified by*
> *Customer No. with Booking No. depend on either*
> *Customer No. alone or on Booking No. alone?*

Yes, all items in that relation are dependent upon Booking No. alone, these are therefore removed to new relation with Booking No. as the key. Note that removing them leaves behind a key only relation of Customer No. with Booking No.

UNF	1NF	2NF	3NF
<u>Customer No.</u>	<u>Customer No.</u>	<u>Customer No.</u>	<u>Customer No.</u>
Customer Name	Customer Name	Customer Name	Customer Name
Veh Cat Code			
Req Bookg Start Date	<u>Customer No.</u>	<u>Customer No.</u>	<u>Booking No.</u>
Req Bookg Finish Date	<u>Booking No.</u>	<u>Booking No.</u>	*Veh Cat Code
Driver Req (Y/N)	Veh Cat Code		*Customer No.
Office No. (Start)	Req Bookg Start Date		Req Bookg Start Date
Office Name (Start)	Req Bookg Finish Date	<u>Booking No.</u>	Req Bookg Finish Date
Office No. (Finish)	Driver Req (Y/N)	Veh Cat Code	Driver Req (Y/N)
Office Name (Finish)	Office No. (Start)	Req Bookg Start Date	*Office No. (Start)
Booking No.	Office No. (Finish)	Req Bookg Finish Date	*Office No. (Finish)
	Office Name (Finish)	Driver Req (Y/N)	
		Office No. (Start)	<u>Office No.</u>
		Office Name (Start)	Office Name
		Office No. (Finish)	
		Office Name (Finish)	

Fig. 6.14

4. Represent the data in Third Normal Form by removing any data items not directly dependent on the key to separate relations Looking at the relation identified by Booking No. we see that there is an inter-data dependency between Office No. and Office Name, which therefore form a separate relation with Office No. as the key. It is not necessary to show two separate relations for the start and finish offices, they can be combined (optimized) together as all offices act as both starts and finishes. However, both Office No. (Start) and Office No. (Finish) must be left in the relation identified by Booking No else information would be lost. These are marked with asterisks in Fig.6.14 to indicate that they are acting as foreign keys.

Remember that we should also look for inter-data dependencies within the compound keys. Asking the questions of the key-only; relation of Customer No. and Booking No.:

> *Given a value for Customer No. is there just one
> possible value for Booking No.?*

No, for one customer there may be several bookings.

> *Given a value for Booking No. is there just one
> possible value for Customer No.?*

Yes, each Booking No. is unique and a booking can be for only one customer.
This means that Customer No. is dependent on Booking No. and therefore the relation becomes:

Booking No.

*Customer No.

Customer No. becomes a foreign key and the relation can be combined with the other relation identified by Booking No. to give the Third Normal Form relations shown in Fig. 6.14. Applying the TNF tests to these relations confirms that they are in Third Normal Form.

Normalization of other Yorkies I/O Descriptions
We describe below the normalization of some of the other data sources identified previously. In order to save space we have combined the I/O Description information about the use and structure of the input or output with the unnormalized form column used for normalization (usually the normalization process would be represented as shown in the Drug Card and Booking Request examples). Only brief comments are made about each normalization.

Daily Departure and Return List (Fig. 6.15) This is a list of the vehicle departures and returns expected at a particular depot on a particular day. This list is printed at the end of the previous day's business in the office and collected by the depot staff before starting work in the morning.

Date of Report has been removed prior to normalization as it is assumed that this will correspond to either Date Booking Starts or Date Booking Ends. No distinction has been made between the departing bookings and the returning bookings as this distinction is implied by the various dates.

There are several optional fields in this report: Driver No., Driver Name, and Agency Name. They are optional because a booked vehicle may be driven by a customer, by a Yorkies driver, or by an agency driver. Some analysts mark optional data items with a preceding 'O'. This can give rise to optional foreign keys e.g. O* Driver No. The results of the analysis of the list are shown in Fig. 6.15.

Normalization of Local Office Vehicle Report (Fig. 6.16) This printed report details all of the vehicles registered at a particular Local Office broken down by their Vehicle Categories.

In this example we have a nested repeating group, this produces the two-item and three-item compound keys shown in First Normal Form in Fig. 6.16. Part key dependencies are removed from both of these compound key relations. The three-item key-only relation has inter-key dependencies and is merged with the relation identified by Veh Reg Mark to produce the Third Normal Form relations.

UNF	1NF	2NF	3NF
Office No.	Office No.	Office No.	Office No.
Office Name	Office Name	Office Name	Office Name
Depot Address	Depot Address	Depot Address	Depot Address
Date of Report			
	Office No.	Office No.	Booking No.
Then a list of Bookings	Booking No.	Booking No.	*Customer No.
separated into	Customer No.		Veh Cat Req
departures and returns.	Customer Name	Booking No.	Date Booking Starts
The following items are	Veh Cat Req	Customer No.	Date Booking Ends
included for each	Date Booking Starts	Customer Name	Driver Req (Y/N)
Booking. Some will be	Date Booking Ends	Veh Cat Req	Veh Reg Mark
blank on the report and	Driver Req (Y/N)	Date Booking Starts	Agency Name
be completed on return	Veh Reg Mark	Date Booking Ends	*Driver No.
or departure.	Agency Name	Driver Req (Y/N)	Start Mileage
	Driver No.	Veh Reg Mark	Finish Mileage
	Driver Name	Agency Name	Date Collected
Booking No.	Start Mileage	Driver No.	Time Collected
Customer No.	Finish Mileage	Driver Name	Date Returned
Customer Name	Date Collected	Start Mileage	Time Returned
Veh Cat Req	Time Collected	Finish Mileage	Return Condition
Date Booking Starts	Date Returned	Date Collected	*Office No. (Start)
Date Booking Ends	Time Returned	Time Collected	*Office No. (Finish)
Driver Req (Y/N)	Return Condition	Date Returned	
Veh Reg Mark	Office No. (Start)	Time Returned	Customer No.
Agency Name	Office No. (Finish)	Return Condition	Customer Name
Driver No.		Office No. (Start)	
Driver Name		Office No. (Finish)	Driver No.
Start Mileage			Driver Name
Finish Mileage			
Date Collected			
Time Collected			
Date Returned			
Time Returned			
Return Condition			
Office No. (Start)			
Office No. (Finish)			

Fig. 6.15

The No. of Vehicles data item can be calculated by counting the occurrences in the Vehicle relation with the particular Local Office ID and the particular Veh Cat Code. Data items whose value can be found by examining the values of other items are known as *derived data items*. These items can often be removed from the normalized relations without loss of information. However, sometimes the value is derived on a particular date and then the data it is derived from is deleted or amended. In this case the derived data item cannot be removed without loss of information. Frequently derived data items are retained to avoid repeated recalculations, though this sacrifices space and risks inconsistent data for improved performance. It is recommended that derived data items are kept until the optimization or Composite Logical Data Design steps and then reconsidered with respect to the whole system's data.

UNF	1NF	2NF	3NF
Local Office ID	Local Office ID	Local Office ID	Local Office ID
Local Office Name	Local Office Name	Local Office Name	Local Office Name
Local Office Address	Local Office Address	Local Office Address	Local Office Address
Depot Address	Depot Address	Depot Address	Depot Address
Then for each Vehicle	Local Office ID	Local Office ID	Local Office ID
Category the following	Veh Cat Code	Veh Cat Code	Veh Cat Code
	Veh Cat Description	No. of Vehicles	No. of Vehicles
Veh Cat Code	No. of Vehicles		
Veh Cat Description		Veh Cat Code	Veh Cat Code
No. of Vehicles		Veh Cat Description	Veh Cat Description
	Local Office ID		
	Veh Cat Code	Local Office ID	Veh Reg Mark
Then a list of vehicles for	Veh Reg Mark	Veh Cat Code	*Local Office ID
each category registered	Make of Veh	Veh Reg Mark	*Veh Cat Code
at that office. The	Model No.		Make of Veh
following items for each	Date of Reg		Model No.
vehicle.	Date of Purchase	Veh Reg Mark	Date of Reg
	End Month Mileage	Make of Veh	Date of Purchase
	Insurance Class	Model No.	End Month Mileage
Veh Reg Mark	Insurance Renewal Date	Date of Reg	Insurance Class
Make of Veh	MOT Date	Date of Purchase	Insurance Renewal Date
Model No.	Condition	End Month Mileage	MOT Date
Date of Reg		Insurance Class	Condition
Date of Purchase		Insurance Renewal Date	
End Month Mileage		MOT Date	
Insurance Class		Condition	
Insurance Renewal Date			
MOT Date			
Condition			

Fig. 6.16

Normalization of invoices and reminders Invoices are produced every two weeks and for active customers several bookings may be included on the same invoice. Normalization of the invoice (Fig. 6.17) is straightforward to Second Normal Form but some interesting points arise at Third Normal Form. Veh Cat Code is dependent on Veh Reg Mark and so is removed to a separate relation. There are several data items associated with the prices for a category which are removed to the relation identified by Veh Cat Code. The Office Name data items are not directly dependent on the Booking No., they depend on Office No. (shown in the normalization of Booking Request). These are therefore replaced by the directly dependent foreign key items of Office No.

UNF	1NF	2NF	3NF
Invoice No.	Invoice No.	Invoice No.	Invoice No.
Invoice Date	Invoice Date	Invoice Date	Invoice Date
Customer No.	Customer No.	Customer No.	*Customer No.
Customer Name	Customer Name	Customer Name	Invoice Amount
Customer Address	Customer Address	Customer Address	
Invoice Amount	Invoice Amount	Invoice Amount	Customer No.
			Customer Name
An invoice may be for	Invoice No.	Invoice No.	Customer Address
several bookings.These	Booking No.	Booking No.	
items are included for	Veh Cat Code		Booking No.
each Booking:	Veh Cat Description		*Invoice No.
	Veh Reg Mark	Booking No.	*Veh Reg Mark
Booking No.	Office Name (Start)	Veh Cat Code	*Office No. (Start)
Veh Cat Code	Date Collected	Veh Cat Description	Date Collected
Veh Cat Description	Time Collected	Veh Reg Mark	Time Collected
Veh Reg Mark	Start Mileage	Office Name (Start)	Start Mileage
Office Name (Start)	Office Name (Finish)	Date Collected	*Office No. (Finish)
Date Collected	Date Returned	Time Collected	Date Returned
Time Collected	Time Returned	Start Mileage	Time Returned
Start Mileage	Finish Mileage	Office Name (Finish)	Finish Mileage
Office Name (Finish)	Return Condition	Date Returned	Return Condition
Date Returned	Damage Charges	Time Returned	Damage Charges
Time Returned	Driver Hours	Finish Mileage	Driver Hours
Finish Mileage	Hourly Driver Price	Return Condition	Driver Cost
Return Condition	Driver Cost	Damage Charges	No. of Days
Damage Charges	No. of Days	Driver Hours	Total Period Cost
Driver Hours	Daily Price	Hourly Driver Price	Total Mileage
Hourly Driver Price	Total Period Cost	Driver Cost	Mileage Cost
Driver Cost	Total Mileage	No. of Days	Total Booking Cost
No. of Days	Price per Mile	Daily Price	
Daily Price	Mileage Cost	Total Period Cost	Veh Reg Mark
Total Period Cost	Total Booking Cost	Total Mileage	*Veh Cat Code
Total Mileage		Price per Mile	
Price per Mile		Mileage Cost	Veh Cat Code
Mileage Cost		Total Booking Cost	Veh Cat Description
Total Booking Cost			Hourly Driver Price
			Daily Price
			Price per Mile

Fig. 6.17

The normalization of the reminder introduces the *composite key*. In order uniquely to
identify each reminder we need a combination of the Invoice No. and the Reminder No.
items because the Reminder No. is just a sequence number for that invoice (e.g. Invoice
No. 34567; Reminder No. 2). The Reminder No. cannot therefore identify a reminder on its
own, it needs the qualifying item of Invoice No. This special case of a compound key
where one or more of the items has no unique significance within the system is known as a
composite key. The data item that is required to make the non-unique item unique is known

as the *qualifying element,* thus Invoice No. is the qualifying element of Reminder No. The SSADM convention is to show composite keys by surrounding each item with brackets and to show the qualifying elements first, as shown in Fig. 6.18.

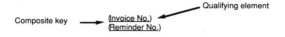

Fig. 6.18

When we go from First to Second Normal Forms we cannot have part key dependency on the non-unique part of the key. So in this example the only part key dependency is on the the qualifying item, Invoice No. (see Fig. 6.19).

UNF	1NF	2NF	3NF
(Invoice No.)	(Invoice No.)	(Invoice No.)	(Invoice No.)
(Reminder No.)	(Reminder No.)	(Reminder No.)	(Reminder No.)
Customer No.	Customer No.	Reminder Date	Reminder Date
Customer Name	Customer Name		
Customer Address	Customer Address	Invoice No.	Invoice No.
Invoice Date	Invoice Date	Customer No.	*Customer No.
Invoice Amount	Invoice Amount	Customer Name	Invoice Date
Reminder Date	Reminder Date	Customer Address	Invoice Amount
		Invoice Date	
		Invoice Amount	Customer No.
			Customer Name
			Customer Address

Fig. 6.19

Optimization of Yorkies relations

We have demonstrated the normalization of the following Yorkies I/O Descriptions: Booking Request and Confirmation (Fig. 6.14); Daily Departure and Return List (Fig. 6.15); Local Office Vehicle Report (Fig. 6.16); Invoice (Fig. 6.17); and Reminder (Fig. 6.19).

As a demonstration we will optimize the relations obtained from the above data sources and in the next section show the full set of optimized relations obtained from all the data sources.

Figure 6.20 shows the three relations that have Booking No. as their primary key. These three relations can therefore be optimized together to form one relation which we shall call the Booking relation. The easiest way of doing this is to take the relation with most data items (i.e. the one from Invoice) and merge the other relations into it. The standard SSADM form for normalization includes a final column, labelled OPT, which should be ticked when a relation has been incorporated into the optimized relation.

Merging the relation from Daily Departure and Return List, we see that there are several items that do not correspond exactly (shown in italics). These should be examined carefully to ensure that they are not synonyms (the same item but with a different name). When merging items of the same name from different relations care should be taken to recognize homonyms (items that have the same name but represent different things; e.g. if we had Date Booking Starts in both the Booking Request and the Invoice, then in the request it would represent the date the vehicle was required and in the Invoice it would represent the date the vehicle was collected—these would not necessarily be the same). The use of a good Data Catalogue, manual or automated, should prove invaluable in avoiding these problems with data item definition.

Booking No.	Booking No.	Booking No.
*Veh Cat Code	*Customer No.	*Invoice No.
*Customer No.	Veh Cat Req	*Veh Reg Mark
Req Booking Start Date	Date Booking Starts	*Office No. (Start)
Req Booking Finish Date	Date Booking Ends	Date Collected
Driver Req (Y/N)	Driver Req (Y/N)	Time Collected
*Office No. (Start)	Veh Reg Mark	Start Mileage
*Office No. (Finish)	Agency Name	*Office No. (Finish)
	* Driver No.	Date Returned
Booking Confirmation	Start Mileage	Time Returned
	Finish Mileage	Finish Mileage
	Date Collected	Return Condition
	Time Collected	Damage Charges
	Date Returned	Driver Hours
	Time Returned	Driver Cost
	Return Condition	No. of Days
	*Office No. (Start)	Total Period Cost
	*Office No. (Finish)	Total Mileage
		Mileage Cost
	Daily Departure and Return List	Total Booking Cost
		Invoice

Fig. 6.20

When we merge in the items from the Booking Confirmation we see that we have several synonyms that can be combined (these are only synonyms if we are dealing with the confirmation rather than the request since it may not be possible to provide what is requested). These are then merged with the most appropriate data item name being used. This gives the Booking relation shown in Fig. 6.21.

When we have produced our optimized relations we must apply the two TNF tests to them to ensure that redundancy has not been introduced during optimization. Applying the first test:

> *Given a value for the key(s) of a TNF relation, is there just one possible value for each data item in that relation?*

Yes, this relation passes the first test.

Applying the second test:

> *Is each data item in a TNF relation directly and wholly dependent on the key(s) of that relation?*

BOOKING

Booking No.
*Invoice No.
*Veh Reg Mark
*Veh Cat Code Req
*Office No. (Start)
*Office No. (Finish)
*Customer No.
*Driver No.
 Agency Name
 Date Collected
 Time Collected
 Start Mileage
 Date Returned
 Time Returned
 Finish Mileage
 Return Condition
 Damage Charges
 Driver Hours
 Driver Cost
 No. of Days
 Total Period Cost
 Total Mileage
 Mileage Cost
 Total Booking Cost
 Veh Cat Code Req
 Date Booking Starts
 Date Booking Ends
 Driver Req (Y/N)

Fig. 6.21

This relation appears to fail the second test in that we seem to have returned the relation to Second Normal Form by including the Veh Cat Code and Customer No. foreign key items. These were directly dependent upon Veh Reg Mark and Invoice No. when the invoice was analysed. However, Customer No. is directly dependent on Booking No. in the early life of a booking as an Invoice No. will not be assigned until later. The Veh Cat Code Req will also be directly dependent in the early life of Booking since the particular vehicle is not assigned until later. It is also possible that the Veh Cat Code Req might be different from the vehicle category actually used, so the removal of this item could mean the loss of valuable information. Therefore this relation does pass the second TNF test.

When the results of this optimization are combined with the relation that results from the analysis of the Vehicle Relocation Booking another typical problem arises. We will merge two relations that have the same key but represent different things. One represents a booking by a customer and the other an internal booking; by merging them we risk losing the information that distinguishes them. Relational theory does allow us to have several different relations that have the same primary key but only if each relation has a different name. However, this is generally not recommended. The usual solution is to merge the two relations but to include a new data item which is simply an indicator to which population the particular occurrence belongs. Thus in this case we have created a new data item called Internal Booking Ind to indicate the internal vehicle relocation bookings.

Local office
Office No.
Office Name
Office Address
Depot Address

Office/cat link
Office No.
Veh Cat Code
No. of Vehicles

Driver
Driver No.
Driver Name

Invoice
Invoice No.
Invoice Date
*Customer No.
Invoice Amount

Customer
Customer No.
Customer Name
Customer Address

Vehicle category
Veh Cat Code
Veh Cat Description
Hourly Driver Price
Daily Price
Price per Mile

Reminder
(Invoice No.)
(Reminder No.)
Reminder Date

Vehicle
Veh Reg Mark
*Office No.
*Veh Cat Code
Make of Veh
Model No.
Date of Reg
Date of Purchase
End Month Mileage
Insurance Class
Insurance Renewal Date
MOT Date
Condition

Fig. 6.22

The optimization of the remaining relations from the data sources previously analysed is straightforward. We have standardized on Office No. as the name of the identifier of a Local Office. The resulting relations are shown in Fig. 6.22.

Applying the TNF tests to these relations shows that they are in Third Normal Form. The full set of Yorkies optimized relations from all the data sources are given in the next section which demonstrates how the set of relations can be displayed as a Logical Data Structure diagram.

SUMMARY

Relational data analysis is a bottom-up technique based on analysing the interrelationships between data items.

The product of relational data analysis is a set of normalized relations (or tables) which minimize redundancy of data and avoid consistency problems.

Each relation is composed of rows and columns:

- Columns map to data items identified in the early stages of SSADM.
- Each occurrence of a row can be uniquely identified by a combination of data items known as the primary key.

Relational data analysis is applied to set of data items, known as data sources, which are usually required system inputs or outputs.

A normalized set of relations is produced from each data source by following these steps:

- Represent the data items in unnormalized form and pick item(s) to act as key.
- Put the data in First Normal Form by removing repeating groups to separate relations.
- Put the data in Second Normal Form by removing part-key dependent items to separate relations.
- Put the data in Third Normal Form by removing items not directly dependent on the key to separate relations.

The above process is often called normalization.

The results of normalization of each data source are combined together by merging all relations with the same key. This process is known as optimization.

After optimization two tests are applied to ensure that the data is in Third Normal Form.

- Given a value for the key(s) of a TNF relation, is there just one possible value for each data item in that relation?
- Is each data item in a TNF relation directly and wholly dependent on the key(s) of that relation?

EXERCISE

Scapegoat Systems
Apply relational data analysis to the staff allocation sheet and the invoice from Scapegoat Systems.

The following end products are required:

1. A normalized set of relations from the analysis of each document .
2. The optimized set of relations.

All information necessary to complete this question is given and implied by the sample documents. However, any assumptions made should be clearly stated.

Staff allocation sheet This indicates which staff are allocated to a project. A staff allocation sheet (Fig. 6.23) is drawn up for each project as the contract is awarded. The No. of Days column relates to the planned number of days that an employee will spend on the project. A project is assigned a unique code and each project will be carried out for a single customer.

Proj Code	3411		**Proj Desc**	New Accounts
Cust No.	3475		**Cust Name**	British Bakers

Staff No.	**Name**	**Grade**	**No. of Days**
34	Bloggs	S.Prog	12
12	Jones	Analyst	3
23	Brown	Manager	9

Proj Code	2356		**Proj Desc**	Betting System
Cust No.	5134		**Cust Name**	Bobs Bookies

Staff No.	**Name**	**Grade**	**No. of Days**
34	Bloggs	S.Prog	3
12	Jones	Analyst	32
45	Williams	Teaboy	12

Fig. 6.23

Invoice Each active customer is invoiced once a month for the work performed by Scapegoat in the previous month. The Start and Finish Dates refer to the overall project but the Man Days column refers only to those worked in that particular month (see Fig. 6.24).

Invoice No:	3412	**Date of Invoice**	23/12/88

From: SCAPEGOAT Systems
Acacia St
London
W14 3RT

Cust No. 3475
Cust Name British Bakers
Address Bread House
Albert Square
London
E12 6TY

Proj Desc	**Start Date**	**Finish Date**	**Man Days**	**Cost**
New Accounts	12/8/88	11/11/88	13	
£13,000				
Delivery System	3/3/88	30/11/88	58	
£42,000				
			Total Cost	£55,000

Fig. 6.24

6.3   Composite Logical Data Design (step 420)

Introduction

In this section we describe how the Composite Logical Data Design is produced by merging the results of relational data analysis derived in step 410 with the Required System Logical Data Structure.

Objectives

There are three major objectives of the step.

1. To produce a detailed data design that will fully meet the user's requirements and is flexible enough to incorporate future requirements.
2. To produce a design that is sufficiently detailed to be converted to a physical database design or file design.
3. To ensure that volumetric information (about the size of data items, the number of the data group, and relationship occurrences) is complete and accurate.

Approach

As described in the introduction to this chapter, this involves the merging of the structure produced by relational data analysis with the Required System Logical Data Structure. The normalized relations, produced in step 410, are groups of data items with primary and foreign key items identified. In order to merge the two representations of the system's data, they must first be put into a comparable form. Thus the set of normalized relations is, by the application of simple rules, represented by a Logical Data Structure type diagram.

The two diagrams are then compared and the differences resolved by reference to the user and to the processing requirements. A merged structure is produced which is known as the Composite Logical Data Design. Differences between the Entity Descriptions and the data content of the normalized relations are resolved in a similar way and merged data group descriptions produced. (The terms entity, relation, and data group are often used interchangeably in SSADM documentation. We only use entity when the Logical Data Structure technique is used, relation when relational data analysis is used, and data group when the Composite Logical Data Design is developed. So entities and relations are merged to form data groups.)

Finally the detailed volumetric information on the data, known as the Logical Design Volumes, are completed. This requires such figures as the number of characters required for each data item, the number of occurrences for each data group and the volumes of relationships. Much of this information will have been collected in earlier stages but additional information may be required on any new data groups or data items identified through relational analysis.

This Composite Logical Data Design and associated supporting documentation are used to specify the detailed enquiry and update processing in stage 5. This specification of the processing validates, in a formal way, that the Composite Logical Data Design can support the processing required of it.

The activities involved in this step are shown diagramatically in Fig. 6.25.

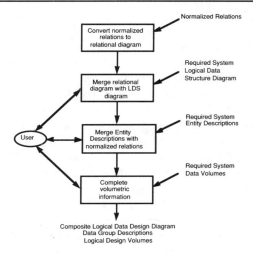

Fig. 6.25

Creating a data structure diagram from normalized relations

In order to compare the results of relational data analysis with the Logical Data Structure derived in stage 2 the normalized relations must be represented in a form that makes comparison easy. It is very simple to represent a set of normalized relations as a data structure diagram using the same conventions as those used in the Logical Data Structure diagram. SSADM prescribes a set of rules for this conversion. However, before applying these rules to the case study we demonstrate how a set of relations can be represented as a data structure.

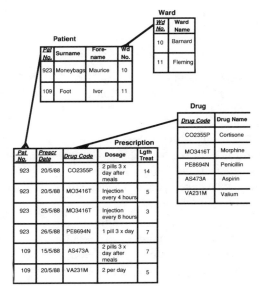

Fig. 6.26

In Fig. 6.26 the four relations identified in Sec. 6.2 from the hospital example are shown. The relationships are shown by crow's feet linking the common keys in each relation. Notice that the 'one' end of the relationship is where the item appears as the sole primary key of the relation. Thus the Ward relation, which has Ward No. as its single primary key, 'owns' the Patient relation, which has Ward No. as a foreign key. Also the Patient relation 'owns' the Prescription relation, which has Patient No. as part of its primary key. This clearly fits in with the data as for one row in the Patient relation (923, Moneybags, etc.) there are four rows in the Prescription relation with the same value of Patient No.

Below we show the conventional SSADM representations for this data: first the set of relations and secondly, the data structure diagram (Fig. 6.27). From these the basic rules for representing a set of relations as a data structure diagram clearly emerge.

1.   Each relation is shown as an entity.

2.   Compound key relations are owned by the relations that have the parts of the compound key as their primary key (thus Patient and Drug own Prescription). If a part of a compound key does not exist as a primary key of another relation then this data item becomes an operational master of the compound key relation (Prescription Date becomes an operational master of Prescription).

3.   Relations that have foreign keys are owned by those relations that have the foreign key as their primary key (Patient is owned by Ward).

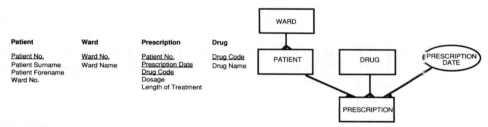

Fig. 6.27

Application to the Yorkies case study

We now demonstrate the representation of the Yorkies normalized relations as a data structure diagram. Figure 6.28 shows the full set of Yorkies optimized relation from relational data analysis.

In Fig. 6.22 we only showed the optimized relations from the data sources analysed in the section; in Fig 6.28 we have shown the full set of relations derived from a complete analysis of all selected data sources.

These will now be developed into a data structure diagram using a systematic approach which follows three simple rules.

1. Relations are shown as entities Each relation is shown as an entity on a Logical Data Structure diagram with the name of the relation inside the box. Often the primary and foreign keys of the relation are also written inside the box—this can help identify which relationships need be drawn.

Local Office
Office No.
Office Name
Office Address
Office Telephone No.
Office Manager Name
Depot Address
Depot Telephone No.

Office/Cat Link
Office No.
Veh Cat Code
No. of Vehicles

Driver
Driver No.
Driver Name
Driver Address
Driver Telephone No.
Driver Comments
Full time/Freelance
*Office No.

Invoice
Invoice No.
Invoice Date
*Customer No.
Invoice Amount

Payment
Payment Identifier
Date of Payment
Amount
Method of Payment
Cheque No.
*Customer No.

Inv/Payt Link
Payment Identifier
 Invoice No.

Licence
Driver No.
Driving Licence Group
Date Test Passed

Customer
Customer No.
Customer Inv Address
Customer Cont Name
Customer Cont Address
Customer Cont Tel No.
Credit Limit

Vehicle Category
Veh Cat Code
Veh Cat Description
Size Code
Max Load Code
Goods Type Codes
Daily Price
Price /Mile
One-way Price /Mile
Hourly Driver Price
Driving Licence Group

Reminder
(Invoice No.)
(Reminder No.)
 Reminder Date

Vehicle
Veh Reg Mark
*Office No.
*Veh Cat Code
Make of Veh
Model No.
Date of Reg
Date of Purchase
End Month Mileage
Insurance Class
Insurance Renewal Date
MOT Date
Condition

Booking
 Booking No.
*Invoice No.
*Veh Reg Mark
*Veh Cat Code Req
*Office No. (Start)
*Office No. (Finish)
*Customer No.
*Driver No.
 Agency Name
 Date Collected
 Time Collected
 Start Mileage
 Date Returned
 Time Returned
 Finish Mileage
 Return Condition
 Damage Charges
 Driver Hours
 Driver Cost
 No. of Days
 Total Period Cost
 Total Mileage
 Mileage Cost
 Total Booking Cost
 Date Booking Starts
 Date Booking Ends
 Date booking Reciv
 Driver Req (Y/N)
 Internal Booking Ind

Fig 6.28

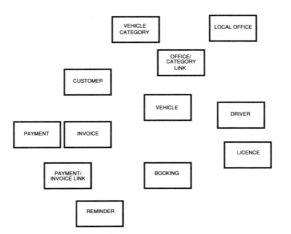

Fig. 6.29

The Logical Data Structure diagram from stage 2 will help in placing the boxes. Another good guideline is to show relations that have large compound keys or many foreign keys at the bottom of the diagram. In Fig. 6.29 we have followed the layout of the Required System Logical Data Structure (Fig. 4.19) to help position the boxes.

2. Compound key relations are owned by the relations which have the parts of the compound key as their primary key (little keys own big keys) Thus any compound key relation is owned by the relations that have the data items contained in the compound key as their primary key. So if a relation, *ABC*, has a compound key, *abc*, then it could be owned by relations *A* (with a key of *a*), *B* (with a key of *b*), and by *C* (with a key of *c*). Alternatively if there were another compound key relation *AB* (with a key of *ab*), then this should own *ABC* rather than relations *A* and *B*. The number of relationships should always be minimized in this way if possible.

If a part of a compound key does not exist as a primary key of another relation then this data item becomes an operational master of the compound key relation. If the data item occurs in any other relations then it should be marked in these with an asterisk to indicate a foreign key. In the case study the compound key of Licence is Driver No. with Driving Licence Group. There is no relation with Driving Licence Group as its key so an operational master is created that 'owns' Licence. Driving Licence Group occurs in the Vehicle Category relation and is therefore marked as a foreign key.

A composite key is a special case of a compound key in that one or more of the items participating in the key have no unique significance in the system. Composite key relations should be owned by the relations that have the unique part of the composite key (the qualifying element) as their primary key. An operational master is *not* created for the non-unique part of the key. In the case study the key of Reminder is a composite key of Invoice No. (the qualifying element) and Reminder No. (the non-unique part). Reminder is therefore 'owned' by Invoice only.

The relationships that have been created under Rule 2 are shown in Fig. 6.30.

Fig. 6.30

3. Relations that have foreign keys are owned by those relations that have the foreign key as their primary key (crow's feet grab the asterisks) A foreign key in a relation indicates that it should be owned by the relation which has that data item as its primary key. Payment has a foreign key item of Customer No., it is therefore owned by the Customer relation which has the primary key of Customer No. Some relations that have many foreign key items can have these items combined into foreign compound keys. For example, the relation *D* which has foreign keys *a, b,* and *c,* should be owned by the relation *AB* (with key *ab*), if it already exists, rather than be separately owned by the *A* and *B* relations. This follows the general principle of minimizing the number of relationships.

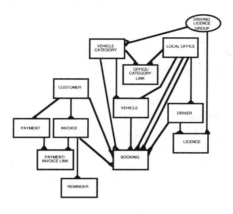

Fig. 6.31

Merging the relational structure with the Logical Data Structure

We now have two data structure diagrams, one derived from the Logical Data Structure technique and one derived from relational data analysis. These diagrams are merged to form one diagram representing the Composite Logical Data Design.

The comparison between the two diagrams, necessary for their merger, will show many differences in structure which need to be resolved with the future users of the system and against the processing requirements. These differences may result from errors in one of the structures; this error detection demonstrates one of the strengths of employing two distinct techniques for data design. Differences will also result from the ways in which the structures are developed; the relational structure will usually be more complex and flexible to future requirements, whereas the Logical Data Structure will usually be simpler and tailored to the specific business requirements identified in systems analysis.

Below we describe the approach to this merging, using the case study. The Required System Logical Data Structure, developed in stage 2, is repeated in Fig. 6.32.

If the two structures are simple, i.e. they have a small number of boxes and relationships, then comparison is easy and differences can be readily identified and resolved. With complex structures a more systematic approach to the comparison is required. Although the Yorkies structures are quite simple we will describe the systematic approach.

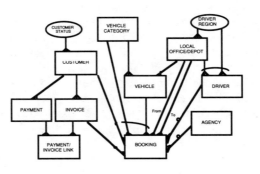

Fig. 6.32

1. Review names of relations and entities

Most relations and entities shown in the two diagrams should match each other exactly. (Often the Logical Data Structure is used to help name the relations in the relational data analysis step.) However, problems can arise when an entity and relation:

- have the same name but are, in fact, different;
- are the same but have different names.

To avoid these misunderstandings a careful review of the names of relations and entities should be carried out first. It may be necessary to compare the data content of the relation with the entity description to resolve the problem.

There is only one such conflict in the Yorkies case study; between the relation Local Office, and the entity Local Office/Depot. As these clearly represent the same thing and as the entity name is judged more appropriate, they can be merged with the name Local Office/Depot.

2. Draw up all entities and relationships that appear in both structures

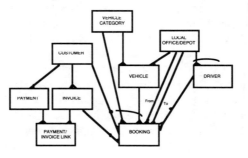

Fig. 6.33

Figure 6.33 shows the common entities and relationships from both diagrams. One apparent difference between the two structures is that the Logical Data Structure diagram shows optional and exclusive relationships which are not shown on the relational structure. These relationships were identified in step 240 and are described in Sec. 4.6. Relational data analysis does not explicitly seek to identify optional or conditional foreign key items which would then appear as optional or exclusive relationships in the relational data structure. These optional and exclusive relationships should be checked against the data analysis working papers to ensure that no conflict arises.

3. Resolve different entities that appear on each structure

Looking just at the entity boxes and their relationships, not the operational masters, several differences emerge:

- Reminder, Licence, and Office/Category Link relations appear on the relational structure.
- An Agency entity appears on the Logical Data Structure.

Each of these differences is resolved against the processing requirement and with the users to decide whether a Composite Logical Data Design data group should be created:

Reminder   After discussion with the users it is decided that, as only three reminders are issued before legal action is started, the Reminder Dates could be held in the Invoice data group. Reminder is not included in the Composite Logical Data Design but the Invoice data group description is amended.

Licence   This was omitted from the Logical Data Structure; the users decide they prefer the extra flexibility offered by its inclusion. This will help in assigning drivers to bookings. Licence is therefore included in the Composite Logical Data Design.

Office/Category Link   This relation was identified by the analysis of the Local Office Vehicle report (Fig. 6.16). The relation shows the number of vehicles of a particular category at a particular office at any time. Further analysis of the data structure shows that this information can be derived from the data items held in Booking and its relationships with Vehicle, Vehicle Category, and Local Office/Depot. The Office/Category Link is therefore omitted from the Composite Logical Data Design.

Agency   This entity was identified as part of the Current System Logical Data Structure. No corresponding relation was identified in step 410 although the Agency Name data item was included in the Booking relation. As other data items were assigned to the Agency Entity Description the users decide it should be included in the Composite Logical Data Design.

4. Resolve different operational masters that appear on each structure

Driving Licence Group   This operational master was created to act as a master to the compound key Licence relation when the relational structure was developed. This is required in the new system to help decide which drivers are licensed to drive which vehicles and is therefore included in the Composite Logical Data Design. Further investigation shows that there are data items associated with the Driving Licence Group so it is shown as a data group.

Driver Region   This operational master was created in the development of the Required System Logical Data Structure as a result of the chosen Business System Option (see Section 4.6). It is included in the Composite Logical Data Design.

Customer Status   This was included in the Required System Logical Data Structure to enable 'off-the-street' customers to be identified. As a high proportion of the customers will be of this status (more than 25 per cent) it was decided that it would not be worthwhile to have direct access on Customer Status but to include it as an item in the Customer data group.

5. Resolve any differences in relationships between the structures

Many of the relationships that appear in one structure and not in the other will have been sorted out by the resolution of differences in the entities and operational masters. However, some differences may still occur. These should be resolved by reference to the processing requirements and, if necessary, to the users.

All differences in relationships in the case study were resolved in the earlier stages of comparison. The completed Composite Logical Data Design diagram is shown in Fig. 6.34.

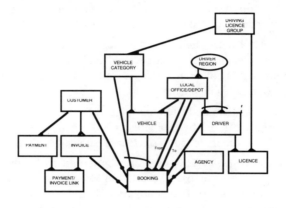

Fig. 6.34

Support documentation to the Composite Logical Data Design

When the Composite Logical Data Design diagram has been produced its support documentation can be finalized. This will require production of Data Group Descriptions from a merger of the Entity Descriptions and of the normalized relations. The volumetric information, required for assessing the amount of storage, is also consolidated from that collected in earlier steps.

Data Group Descriptions

These will be used to define the detailed data content of the new system and will be used for physical data definition of records, files, tables, and so on in stage 6, physical design, and in the construction phases of the development. The production of these Data Group Descriptions requires several activities.

- Extending the Entity Descriptions developed in steps 130 and 240 to include the data content identified by relational data analysis.
- Developing Data Group Descriptions for any new data groups identified in stage 4.
- Ensuring that the data groups identified by logical data structuring alone have their primary and foreign keys correctly defined.
- Ensuring that the Data Catalogue (described in Sec. 3.4) is complete, new entries may be required for items identified in stage 4.

At the end of the step every data group should be fully documented as follows:

- A unique name of the data group.
- A description of what the data group represents.
- The names of the data groups that own and are owned by the data group.
- A list of the data items included in the data group showing the primary keys, foreign keys, and any conditions or role names under which the items are included or excluded from the data group.
- Volumetric information about the space requirement for each occurrence and the number of occurrences (although this may be held as Logical Design Volumes—see below).

At the end of the step each data item should be fully documented as follows:

- A unique name of the data item.
- A description of what the data item represents.
- Format and range information.
- Any validation criteria that may be applied to check that the data entered is correct.
- Number of characters required and whether of fixed or variable length (if variable the average and maximum lengths).

Ideally also every relationship on the Composite Logical Data Design should have a description of the meaning of the relationship and the conditions under which it exists or precludes other relationships.

This documentation of data groups, data items, and relationships is best held in a computerized data dictionary package. This could be part of an integrated computer support tool to SSADM or could be the data dictionary employed by the database management system to be used for the development.

Volumetric information
Detailed volumetric information is required to determine the amount of disc storage required and to time the transactions performed by the system. In this step we ensure that all the volumetric information on the data has been collected and fully documented. Much of this information will have been collected in the earlier stages of the project.

One way of presenting all this information in a convenient way is to use the SSADM Logical Design Volumes form. This shows the space requirement for each data group and the number of occurrences participating in each relationship. An example of the form is shown in Fig. 6.35.

This form is a summary of all the volumetric information we hold about the data. More detailed information may be documented with the data group, data item and relationship descriptions which may be held in a computerized form. Below we explain the meaning and purpose of the various areas of the form.

Data group reference and name A reference number is often given to entities and data groups. In this simple example we have omitted them and given only the data group name.

Data group length This is the sum of the lengths of the individual data items that form the data group. If there are variable length items then the maximum lengths are usually summed.

Master data group The names of the data groups owning the one shown in the Data group name column are written here. So as Licence is owned by both Driving Licence Group and by Driver these are shown linked to Licence by a '<'.

Dependent volumes These show the number of occurrences of the detail data group for one occurrence of the master. Thus there are, on average, 41 vehicles belonging to a vehicle category and 50 vehicles registered at a particular local office/depot. The maximum number of vehicles belong to a category is 150. The comments column can be used to explain any of the figures associated with the data group.

Volatility per... This shows how often the data is updated over a given period (per week, per month, etc.). The frequency of insertions, deletions, and amendments can be shown here.

Number of occurrences This shows the average total number of occurrences of the data group.

Logical file size This shows the total space that would be occupied by the average number of occurrences of the data group and is given by 'Lgth' multiplied by 'No. Occ.' Note that this is not the actual space that would be taken up on disk as no allowance has been made for overheads such as page management, indices, overflow, etc.

Logical Design Volumes

System: Yorkies			Date:				Author:					
Data grp ref.	Data group name	Lgth	Master data group	Dependent volumes			Vol'ty per...			No. Occ.	Lgcl file size
				Avg	Max	Comments	Ins	Del	Amd		
	Local Office/Depot	155	Driver Region	4	6					50	7750
	Driving Licence Group	25								5	260
	Licence	20	< Driving Licence Group / Driver	840 / 2	2600 / 5					4200	84000
	Driver Region	10								14	140
	Driver	95	< Driver Region / Local Office/Depot	150 / 10	250 / 15					2600	247000
	Vehicle Category	90	Driving Licence Group	12	18					61	5490
	Vehicle	110	< Vehicle Category / Local Office/Depot	41 / 50	150 / 75					2500	275000
	Agency	45								12	540
	Booking	140	Local Office/Dep(X2) / Driver / Agency / Vehicle / Vehicle Category / Customer / Invoice	2x1050 / 12 / 410 / 18 / 115 / 17 / 2.5	2x1600 / 70 / 800 / 40 / 180 / 60 / 10	Bookings kept for 6 months before archive Bookings taken 1 month ahead. No. of Bookings will increase—25% p.a?				52500	7350000
	Customer	124				No. Customers increase				3000	372000
	Payment	43	Customer	27	50	at 40% p.a.?				81000	3321000
	Invoice	57	Customer	28	50	Increases expected				84000	4788000
	Payment/Invoice Link	16	< Payment / Invoice	1 / 1	5 / 5	35% p.a.?				8250	1320000

Fig. 6.35

Another good and highly visual way of representing the summary volumetric information is to write it on a copy of the Composite Logical Data Design diagram. The numbers of occurrences and the lengths of each data group can be written inside the boxes. The volumes of relationships can be written on the relationship lines.

SUMMARY

- The Logical Data Structure and the relations derived from relational data analysis are combined to form the Composite Logical Data Design.
- The relations are represented as a data structure diagram to aid comparison with the Logical Data Structure.
- The two diagrams are merged and differences resolved with reference to the processing requirements and to the users.
- The Entity Descriptions are extended to show the full data content defined by the relations.
- Any remaining documentation of the systems data, such as Data Group Descriptions, Data Item Descriptions, and Relationship Descriptions, is completed.
- All volumetric information on the data is consolidated.

EXERCISE

Scapegoat Systems

1. Develop a data structure diagram to represent the optimized relations obtained in the relational data analysis exercise in Sec. 6.2.
2. Compare this with the Required System Logical Data Structure, developed in Sec. 4.6, to give the Composite Logical Data Design.

7. Process design (stage 5)

7.1   Introduction

Purpose and overview of stage 5

In stage 5, using the Composite Logical Data Design as a basis, the required processing is specified in detail on Process Outlines. The physical file design and file handler (or database design and DBMS) are not taken into account until stage 6, physical design.

Attitudes are mixed as to whether it is worth producing such a mass of documentation which is logical (i.e. hardware and software independent) when, for the majority of projects, the hardware and software that will be used for the construction are known at this stage. Many people would suggest that the Process Outlines should be written in the target language for direct implementation. There is no clear answer to this. Some experience has shown that attempting to produce implementation-specific documentation at this stage is too big a jump. By specifying the processing in logical terms first, there is a more gradual progression towards the greater level of detail needed for implementation-specific specifications. Also, the logical specifications are then available for possible future changes to the system. However, if fourth-generation development tools are being used, much of the detail of Process Outlines is unnecessary and this is considered a good point at which to start directly prototyping. These questions have not been satisfactorily resolved and it seems likely that this is the stage of SSADM that will undergo the most significant changes in the next few versions of the method.

Inputs and outputs of the stage

The two types of Process Outline produced in this stage are Logical Update and Logical Enquiry. A separate Logical Update Process Outline is produced for each event or sub-event identified on the Entity Life Histories. First, the Entity Life Histories and Logical Dialogue Outlines are reviewed to ensure that differences between the Logical Data Structure and the Composite Logical Data Design have been reflected there and to ensure that the requirements of the user are still faithfully reflected.

A Logical Enquiry Process Outline is produced for each enquiry function on the Retrievals Catalogue.

The inputs to the Process Outlines are the Composite Logical Data Design, Logical Dialogue Outlines, the I/O Descriptions for inputs to and outputs from the processes and the Elementary Function Descriptions that support the Required Data Flow Diagrams. For the Logical Enquiry Process Outlines, the Retrievals Catalogue is also required and for the Logical Update Process Outlines, the Entity Life Histories and the Event Catalogue are required.

The complete set of Process Outlines is used to validate the Composite Logical Data Design. All of the documentation produced in this stage is reviewed in detail at the end of the stage.

Outputs from the stage include a complete set of Logical Process Outlines (Update and Enquiry) completely consistent with all other SSADM documentation. If changes have been required to the Entity Life Histories, Event Catalogue, Dialogue Outlines, and Composite Logical Data Design, the amended versions of these documents may also be considered an output of this stage.

7.2   Revision of Entity Life Histories (step 520a)

Introduction

Revisions to the Entity Life Histories are necessary after the completion of the Composite Logical Data Design as the logical data groups may not map directly onto the entities that they replace. Also, it is necessary to confirm that the requirements of the user are still accurately reflected in the Entity Life Histories before specifying the detailed processing on the Process Outlines.

The Required System Logical Data Structure may have undergone some significant changes when expanded to become the Composite Logical Data Design. For example, areas of the required data that may have been completely overlooked in the analysis are picked up by relational data analysis. Similarly, entities might be split into several logical data groups, entities might be merged into one logical data group, relationships between logical data groups might be different from the relationships between the corresponding entities, or new data items might be added to the logical data groups. All of these changes affect the Entity Life Histories. Before it is possible to continue with the logical process design, the Entity Life Histories must be brought into line with the Composite Logical Data Design.

In the Yorkies system, the data groups Driving Licence Group and Licence have been introduced in the Composite Logical Data Design; there were no corresponding entities in the Required System Logical Data Structure. Obviously, if a completely new data group is added to the Composite Logical Data Design, an Entity Life History for it will not already exist. To be sure that the logical design is still correct and consistent, Entity Life Histories are drawn for the new additions. The original ELH Matrix is used to check whether any of the events previously identified will affect the new data groups. Also, any new events that are identified at this stage are checked against the existing data groups to see if they might be affected. An extract from the Yorkies ELH Matrix is shown with the new data groups in Fig. 7.1.

Do any of the existing events affect either of the new entities? Licence is a detail of Driver so it is likely that it will be affected by some of the same events. On this part of the matrix, the events that affect Driver are New Driver and Driver Departure. It is likely that when a driver joins Yorkies, he or she will already possess a licence of some kind! Thus, the event Driver Joins will create the relevant occurrences of Licence as well as creating the Driver data group. Similarly, when the driver leaves, the record of his or her licences will be deleted at the same time as the other driver data.

	BOOKING REQUEST	NEW CUSTOMER	DRIVER ASSIGNED	AGENCY ALLOCATED	VEHICLE DEPARTURE	VEHICLE RELOCATION	VEHICLE RETURN	BOOKING CONFIRMED	PAYMENT DUE	PAYMENT RECEIVED	CUSTOMER CHANGE	NEW DRIVER	DRIVER DEPARTURE
CUSTOMER	I	I									M		
INVOICE							I	M					
PAYMENT	I									I			
PAYMENT/INVOICE										I/M			
BOOKING	I		M	M	M	I	M	M					
VEHICLE						M							
VEHICLE CATEGORY													
DRIVER												I	D
AGENCY													
DRIVING LICENCE GROUP													
LICENCE													

Fig. 7.1

The Driving Licence Group data group is not affected by any of the events currently on the ELH Matrix. The events which will affect this data group are identified as New Driving Licence Group and Driving Licence Group Obsolete. This latter event is likely to affect Licence also. The matrix is now updated to show these new events and effects, as shown in Fig. 7.2.

	VEHICLE DEPARTURE	VEHICLE RELOCATION	VEHICLE RETURN	BOOKING CONFIRMED	PAYMENT DUE	PAYMENT RECEIVED	CUSTOMER CHANGE	NEW DRIVER	DRIVER DEPARTURE	NEW DRIVING LICENCE GROUP	DRIVING LICENCE GROUP OBSOLETE
CUSTOMER							M				
INVOICE			I	M							
PAYMENT						I					
PAYMENT/INVOICE						I/M					
BOOKING	M	I	M	M							
VEHICLE		M									
VEHICLE CATEGORY											
DRIVER								I	D		
AGENCY											
DRIVING LICENCE GROUP										I	D
LICENCE								I	D		D

Fig. 7.2

The Entity Life Histories of the two new logical data groups can now be drawn as before, possibly identifying new events which will be reflected back onto the matrix. The final Entity Life Histories are shown in Fig. 7.3.

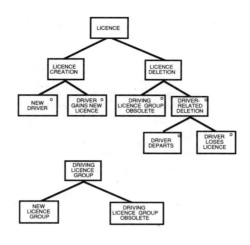

Fig. 7.3

A less obvious change during the data design stage will also result in changes to an Entity Life History: some new data items have been added to the Invoice data group. The Invoice Entity Life History as it was after stage 2 is shown in Fig. 7.4.

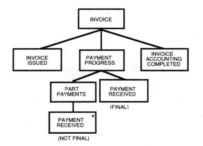

Fig. 7.4

The Entity Life History of the Invoice entity must now be reviewed to ensure that the new data items are updated by events. The new data items are:

- Reminder 1 Date;
- Reminder 2 Date;
- Reminder 3 Date;
- Date Legal Action Taken.

From these data items, we can deduce that new events will be needed:

- First Reminder sent;
- Second Reminder sent;
- Third Reminder sent;
- Legal Action taken.

The first three will only affect the Invoice data group, but the final one will affect the Customer data group as well by setting the credit limit to a 'suspended' value.

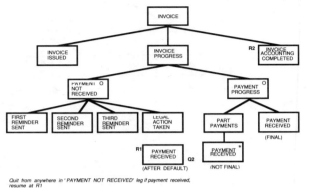

Fig. 7.5

The Invoice Entity Life History is amended to cater for the new events as shown in Fig. 7.5. This Entity Life History illustrates a particular use of the 'quit and resume' convention which is used when events can affect the ELH in no predictable place. The sequence of the three reminders followed by legal action being taken can be stopped at any point if a payment is received from the customer. This means that during this sequence, the receipt of a payment could occur at random. Rather than place the 'Q1' at a specific point, then, this is replaced by a sentence 'Quit from anywhere in Payment Not Received leg if payment received, resume at R1'. The resume point is at an effect that is off the structure as it does not belong at any specific point on the structure because it can happen at random. This is yet another occurrence of the event Payment Received in this Entity Life History. This only happens once as the issue of reminders requires the customer to settle the invoice in full with a single payment. Another quit and resume is added to continue the sequence at R2 so that the invoice is deleted by the invoice accounting completion. This might be a little clearer with the addition of state indicators as in Fig. 7.6.

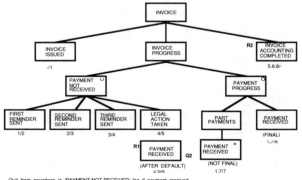

Fig. 7.6

As the 'Valid Previous' values under the last box show, the invoice accounting is done after one of three possibilities has occurred:

- Legal action has been taken (5).
- Payment is received after default (6).
- Payment is received to pay off the invoice in the normal way (8).

Checking the requirements

At this stage of the project it is worth checking again with the users that their requirements are accurately reflected in the logical design. The introduction of the new events in the Invoice Entity Life History in the Yorkies system, for example, may not reflect the way in which the accounts section of Yorkies wishes to handle the chasing of unpaid invoices. It is necessary to be sure that the users are content with the constraints reflected in the Entity Life Histories. Are the specific sequences always going to be true or will there be a need to override them from time to time and in what circumstances? Will the users require more constraints than shown in the Entity Life History? Do the Logical Dialogue Outlines still accurately reflect the way in which the users wish to conduct the dialogues? These factors ought to be settled before the Process Outlines are produced, if at all possible, because there is a massive amount of effort about to be expended and it needs as firm a base to build upon as possible.

SUMMARY

- Entity Life Histories should be reviewed after the production of the Composite Logical Data Design.
- New entities are added to the ELH Matrix, and new Entity Life Histories produced for them.
- If any Entity Descriptions have changed then the corresponding Entity Life Histories may need to be changed to reflect new events affecting the new data items.
- The user's requirements are checked to ensure the Entity Life Histories accurately reflect what is required.

7.3   Definition of logical processing (steps 510 and 520b)

Introduction

Process Outlines are documents that describe the detail of the operations that the system will perform either in response to an event (Logical Update Process Outlines) or an enquiry/retrieval (Logical Enquiry Process Outline). To understand how Update Process Outlines relate to Entity Life Histories, consider the ELH (entity/event) Matrix that was used to start the Entity Life Histories. This is repeated in Fig. 7.7. The events are along the horizontal axis and the entities are along the vertical axis. If you select an entity and look along the row belonging to that entity, you will see all the events that affect the entity. This is the Entity Life History view. If you select one of the events and look down the column belonging to that event, you will see all the entities that are affected by that event. This is the Process Outline view.

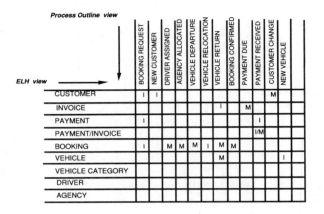

Fig. 7.7

The Process Outlines are traditionally logical documents that do not take account of the target system. Also, although they can be used to show in what order the entities (or, more correctly, data groups) should be accessed to achieve the correct update or enquiry, they do not constitute a process (or program) design. SSADM was originally designed to be used as a basis for the use of program design methods, principally the government standard program design method SDM (Structured Design Method) which is based upon Jackson Structured Programming (JSP). As such, the logical process design of stage 5 should not prevent the later use of these methods to design well-structured programs by dictating a design at this stage. In Chapter 8, there is a discussion of how the use of fourth-generation development tools may influence the Process Outlines and later program specifications.

The reasons for specifying the processing in logical terms before respecifying it in physical terms are principally:

1. The detail of the final specification can be built up step by step. The underlying process design is specified before the extra detail specific to an implementation needs to be addressed.
2. The logical design can be used as a basis for the assessment of hardware and software options.
3. The logical specification is a basis for tuning the physical data design. The principal consideration in the design of the physical database or file structure is whether or not the processing can be performed within the limits specified by the user. This can be ascertained by paper timings of the accesses specified on the Process Outlines.

Process Outlines are used as the basis for the specifications that are given to programmers. In stage 6 (physical design), they are re-expressed in physical terms relating to the physical data design and the database management system (or equivalent) so that they can be understood by the implementers.

The order in which operations are written on the Process Outlines is determined by reference to the structure of the data accessed and any sequences specified on the Logical Dialogue Outlines. It is determined by the structure of the data because the order in which the logical data groups (entities) are accessed through the Composite Logical Data Design will dictate the order in which the processing of each data group is performed. This is illustrated here by a simple example. Figure 7.8 shows a portion of a Composite Logical Data Design containing the data groups A, B, C, and D.

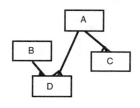

Fig. 7.8

A commonly required type of enquiry might need to access all of the B's related to a particular occurrence of A. (This might be all the customers related to a particular bank account, all the drivers who are booked to drive a particular vehicle, or any other similar enquiry where a many-to-many relationship exists between two entities.) The access path would start with A, find all the related occurrences of D, and, for each occurrence of D, find the related occurrence of B. This would retrieve the information necessary to satisfy the enquiry by accessing the entities in the order: A, D, B. This access path is shown on the diagram by the use of arrows as shown in Fig. 7.9.

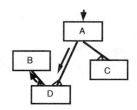

Fig. 7.9

The Logical Enquiry Process Outline for this enquiry would be specified in the order: access A; access D; access B.

For each of these accesses, one or more operations can be specified to describe exactly what the system must do once it has accessed the data groups. This includes validation checks and interaction with the user.

Building up the Process Outlines helps to validate the Composite Logical Data Design. Relationships between the logical data groups must allow the sequence of accesses required by each process.

Each Logical Update Process Outline describes the processing to be done in response to an event. The Process Outlines validate the Entity Life Histories by highlighting possible effects on other logical data groups (entities) on the access path which were not previously considered.

Inputs to Process Outlines

The documents to be assembled as an input to Process Outlines are:

- I/O Descriptions.
- The Composite Logical Data Design.
- Logical Dialogue Outlines.
- Elementary Function Descriptions.

Enquiry only:

- Retrievals Catalogue.

Update only:

- Entity Life Histories.
- The Event Catalogue.
- ELH Error Handling Narrative.

Determining the accessing requirements

It is useful to have a separate copy of the Composite Logical Data Design for each Process Outline produced so that the access path can be clearly sketched out as shown in Fig. 7.9. If a Logical Dialogue Outline is available, this will help to identify the sequence in which the logical data groups will be required.

In the Yorkies system, we have already considered in detail the events that affect the Booking entity (see Sec. 4.7). Here, one of the events is taken from that Entity Life History and the Logical Update Process Outline is built up to support it. The event Receipt of Booking Request not only affects the Booking entity, it also affects the Customer entity (if the booking is from an 'off-the-street' customer, then a Customer entity must be created at the same time as the Booking entity). So, the event Receipt of Booking Request has two effects: one upon the Booking entity and one upon the Customer entity. The Process Outline will show accesses to these two entities. Also, there may be other entities that must be accessed for validation before the updates. (Note that the Entity Life Histories only show whatever triggers updates to the entities and do not reflect enquiries.)

We must decide upon the access path before going on to specify the update processing that is required.

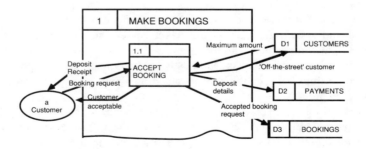

Fig. 7.10

It will help to look at the Data Flow Diagram again. This is shown in Fig. 7.10. The Data Flow Diagram shows a data flow entering process 1.1 from data store D1: Customer labelled 'Maximum amount'. This represents the checking of the customer's credit limit before a booking is accepted. There are three data flows entering data stores (representing updates). As well as the updates to the Bookings and Customers described before, the Payments data store is also updated with details of any deposit made by the customer. The Payments data store represents the Payment entity on the Logical Data Structure, so we now know that the Receipt of Booking Request firstly results in the customer's credit limit being checked, followed by up to three updates being performed on the Customer, Booking, and Payment entities.

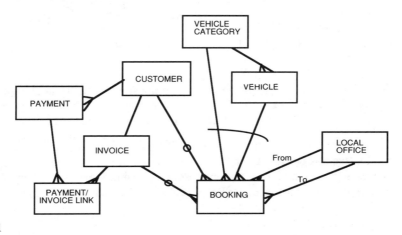

Fig. 7.11

The access path through the Composite Logical Data Design must now be decided. Fig. 7.11 shows a portion of the Yorkies Composite Logical Data Design. Here, the entities that we know to be involved in the 'Receipt of Booking Request' processing are highlighted. The accesses will be required thus:

1. First, the Customer entity is read to make sure a record of the customer exists and that this booking does not exceed the credit limit previously set. If it is an 'off-the-street' customer, a new customer entity occurrence is created. The Customer entity is the 'entry point' for this event.
2. If the customer check is OK, a Booking occurrence can now be created.
3. Next, as the booking will be for a specific category of vehicle, we must access the Vehicle Category entity to check that the one specified on the booking request is correct. The booking is linked to a Vehicle Category occurrence only at this point: the exact vehicle availability is checked later when the booking will be linked to a Vehicle entity.
4. The booking is assigned to a Yorkies local office (or two different offices if it is a one-way hire).
5. Finally, any deposit that has been made will cause the creation of a Payment occurrence.

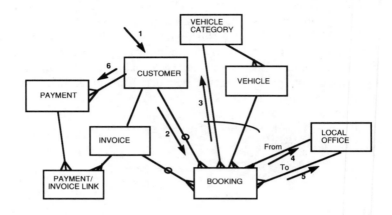

Fig. 7.12

The access path is sketched out on the Composite Logical Data Design in Fig. 7.12. Here, the order in which the accesses are done is marked on each arrow so the access path can be easily traced. It is now a relatively straightforward matter to complete the Process Outline for this event.

Completing the Process Outline forms

A Logical Update Process Outline form is shown in Fig. 7.13. The use of this form will be demonstrated for the Booking Request Received event.

Logical Update Process Outline												
System:		Date:		Author:			Event:					

| Batch / On-line | **Access Type:** (Acc Typ) | I - Insert D - Delete
M - Modify L+ - add Link
R - Read   L- - remove Link | **Read Path :** (Rd Pth) | Dir - Direct
Seq - Sequential
Det - via Detail
Mast - via Master |

Op No.	Operation Description	I/O Ref	Acc No.	Entity	State Ind		Acc Typ	Rd Pth	Access Via	No. Acc	Data Items
					Val Pr	Set					

Fig. 7.13

We have already determined that the entities that are to be accessed are:

1. Customer.
2. Booking.
3. Vehicle Category.
4,5. Local Office.
6. Payment.

The entities, together with their Access No. have been added to the form in Fig. 7.14.

Logical Update Process Outline												
System:		**Date:**		**Author:**			**Event:** Receipt of Booking Request					
Batch / On-line	**Access Type:** (Acc Typ)		I - Insert D - Delete M - Modify L+ - add Link R - Read L- - remove Link				**Read Path :** (Rd Pth)	Dir - Direct Seq - Sequential Det - via Detail Mast - via Master				
Op No.	Operation Description		I/O Ref	Acc No.	Entity	State Ind		Acc Typ	Rd Pth	Access Via	No. Acc	Data Items
						Val Pr	Set					
				1	Customer							
				2	Booking							
				3	Vehicle Category							
				4	Local Office							
				5	Local Office							
				6	Payment							

Fig. 7.14

The following sections describe what each of the columns on this form is for and demonstrates how they are completed for this event.

Op No. (operation number)

This is a system-wide identifier for an individual operation on the Process Outline form. If there is a Logical Dialogue Outline corresponding to this Process Outline, the operation number is used to cross-reference the dialogue steps with the operations on the Process Outline form. If an operation description appears on more than one Process Outline, the operation number will cross-reference each occurrence of the operation to the Common Processing Catalogue where the operation will be described just once. The operations have been added to the Process Outline in Fig. 7.15

Operation description

The operations to be performed are described here briefly in plain English (although it is often preferred practice to complete this in structured English).

Logical Update Process Outline

System:		Date:		Author:		Event: Receipt of Booking Request

Batch / On-line	**Access Type:** (Acc Typ)	I - Insert D - Delete M - Modify L+ - add Link R - Read L- - remove Link	**Read Path :** (Rd Pth)	Dir - Direct Seq - Sequential Det - via Detail Mast - via Master

Op No.	Operation Description	I/O Ref	Acc No.	Entity	State Ind		Acc Typ	Rd Pth	Access Via	No. Acc	Data Items
					Val Pr	Set					
O1 O2 O3	Check Customer exists. Display Credit Limit and accept confirmation to proceed. If 'off-the-street', create Customer		1	Customer							
O4	Create Booking for Customer		2	Booking							
O5	Check if requested Vehicle Category exists. If yes, link Booking to Vehicle Cat		3	Vehicle Category							
O6	Link to appropriate Local Office, using 'From' relationship		4	Local Office							
O7	If one-way hire, link to other Local Office using 'To' relationship		5	Local Office							
O8	If a deposit made, create Payment		6	Payment							

Fig. 7.15

You may notice that some lines have been drawn across the Process Outline. These are entirely optional but are a useful way of separating the operations to do with a particular effect from those belonging to another. Also, as there are so many columns to try to match up across the page, the lines help prevent the eye from 'jumping'!

I/O Description

System :	Yorkies	Sub-system :		Author :		Date :

Name:	Booking Request	Ref No.:	a-1	Type:	Form / screen

Description: Customers send in a Booking Request Form or may give the same information over the telephone. In either case the information is entered into the system on-line to create the booking.

Data items	Format	Length	Comments
Customer No. Customer name Vehicle Category Code			
			The next group of data items may be repeated if there are several Bookings
Required Booking start date Required Booking finish date Driver requirement (Y/N) Office No. (Start) Office name (Start) Office No. (Finish) Office name (Finish)			{ only inserted if { one-way hire

Fig. 7.16

I/O Ref (input/output reference)

This refers to screens, reports, messages, and so on that are input to the process or that are output by the processing. The event Receipt of Booking Request will be implemented as an on-line dialogue and will involve a number of screens. These screens will have been specified on I/O Descriptions at the end of stage 2 or stage 3. The I/O Description to put in the booking request was described in Sec. 4.5. This is repeated in Fig. 7.16. It is necessary to refer to this I/O Description in the appropriate place on the Process Outline.

The I/O Ref column can also be used to refer to messages, especially error messages. For example, the fifth operation described on the form (O5) states: 'Check Vehicle Category requested exists'. If it does not exist, we may wish to specify that the system outputs error message E1: 'Vehicle Category does not exist'. Figure 7.17 shows the Process Outline with the I/O Ref column completed.

For more complex errors, this column can be used to refer to another Process Outline that describes some operations to be performed on the detection of an error.

Logical Update Process Outline

System: Date: Author: Event: Receipt of Booking Request

Access Type: (Acc Typ)   Batch / On-line
- I - Insert   D - Delete
- M - Modify   L+ - add Link
- R - Read   L- - remove Link

Read Path : (Rd Pth)
- Dir - Direct
- Seq - Sequential
- Det - via Detail
- Mast - via Master

Op No.	Operation Description	I/O Ref	Acc No.	Entity	State Ind Val Pr	State Ind Set	Acc Typ	Rd Pth	Access Via	No. Acc	Data Items
O1 O2 O3	Check Customer exists. Display Credit Limit and accept confirmation to proceed. If 'off-the-street', create Customer	a-1	1	Customer							
O4	Create Booking for Customer		2	Booking							
O5	Check if requested Vehicle Category exists. If yes, link Booking to Vehicle Cat	E1	3	Vehicle Category							
O6	Link to appropriate Local Office, using 'From' relationship		4	Local Office							
O7	If one-way hire, link to other Local Office using 'To' relationship		5	Local Office							
O8	If a deposit made, create Payment		6	Payment							

Fig. 7.17

Acc No. (access number)

This is an identifier that is internal to the Process Outline, and may have significance as a sequential count of the number of the access on the Composite Logical Data Design. Where there are conditional branches in the processing, the access number may be repeated to indicate more clearly that some of the operations are alternatives to other operations. These were added in Fig. 7.14.

Entity

Here, the Composite Logical Data Design logical data group being accessed is reflected. This might also include an entity identifier. This column was completed in Fig. 7.14.

State indicators

Val Prev S I (valid previous state indicators)   The range of 'valid previous' state indicator values is placed here. For the Update Process Outlines, these are taken directly from the Entity Life Histories. The completion of this column often removes the need for lengthy validation descriptions; the valid previous values define the state the entities are in before they can validly be affected by this event. For example, validation such as 'A driver cannot be allocated to a booking until the booking has been confirmed' can be replaced by examining the state indicators: if the state indicator is any value other than that to which the Booking Confirmed event set it, the event Driver Allocated cannot be allowed to happen.

By examining the appropriate Entity Life Histories (see Sec. 4.7), we can see that this event causes the creation of Booking, Customer, and Payment occurrences. Therefore, in each of these three effects, the valid previous state indicator will be 'null'.

For Enquiry Process Outlines, this column can be used to denote the state indicator values of logical data groups to be selected for this enquiry.

Set To S I (set to state indicators: LUPO only)   This is the single value that an updating operation will set the state indicator to. These are taken directly from the Entity Life History.

The state indicators are added to the Process Outline for 'Receipt of Booking Request' in Fig. 7.18.

Acc Typ (access type: LUPO only)

This refers to Update Process Outlines only as all accesses on a Logical Enquiry Process Outline will be of the same type, Read. For updating, the possible values entered in this column are:

I   *Insert*: the logical data group is being created.

M   *Modify*: data items within the logical data group are changed. To modify an logical data group, a previous operation must have been of Access Type 'R'.

R   *Read:* the logical data group is read before any operations are performed on it.

D   *Delete:* specified logical data group occurrences are deleted

L+   *Add to link path:* an operation has created a relationship. This will always accompany the Insert access as the creation of a logical data group will always create its relationships. The Modify access may also create relationships.

L-   *Remove from link path:* an operation has deleted a relationship. This will always accompany the Delete access as the deletion of a logical data group will always delete its relationships. The Modify access may also delete relationships.

Logical Update Process Outline

System:	Date:	Author:	Event: Receipt of Booking Request

Access Type: (Acc Typ)   Batch / On-line

I - Insert   D - Delete
M - Modify   L+ - add Link
R - Read   L- - remove Link

Read Path: (Rd Pth)   Dir - Direct   Seq - Sequential   Det - via Detail   Mast - via Master

Op No.	Operation Description	I/O Ref	Acc No.	Entity	State Ind Val Pr	State Ind Set	Acc Typ	Rd Pth	Access Via	No. Acc	Data Items
O1 O2 O3	Check Customer exists. Display Credit Limit and accept confirmation to proceed. If 'off-the-street', create Customer	a-1	1	Customer	–	1					
O4	Create Booking for Customer		2	Booking	–	1					
O5	Check if requested Vehicle Category exists. If yes, link Booking to Vehicle Cat	E1	3	Vehicle Category							
O6	Link to appropriate Local Office, using 'From' relationship		4	Local Office							
O7	If one-way hire, link to other Local Office using 'To' relationship		5	Local Office							
O8	If a deposit made, create Payment		6	Payment	–	1					

Fig. 7.18

Logical Update Process Outline

System:	Date:	Author:	Event: Receipt of Booking Request

Access Type: (Acc Typ)   Batch / On-line

I - Insert   D - Delete
M - Modify   L+ - add Link
R - Read   L- - remove Link

Read Path: (Rd Pth)   Dir - Direct   Seq - Sequential   Det - via Detail   Mast - via Master

Op No.	Operation Description	I/O Ref	Acc No.	Entity	State Ind Val Pr	State Ind Set	Acc Typ	Rd Pth	Access Via	No. Acc	Data Items
O1 O2 O3	Check Customer exists. Display Credit Limit and accept confirmation to proceed. If 'off-the-street', create Customer	a-1	1	Customer	–	1	R I				
O4	Create Booking for Customer		2	Booking	–	1	I, L+				
O5	Check if requested Vehicle Category exists. If yes, link Booking to Vehicle Cat	E1	3	Vehicle Category			R L+				
O6	Link to appropriate Local Office, using 'From' relationship		4	Local Office			L+				
O7	If one-way hire, link to other Local Office using 'To' relationship		5	Local Office			L+				
O8	If a deposit made, create Payment		6	Payment	–	1	I L+				

Fig. 7.19

On the Process Outline for the Booking Request, the Customer is initially read and possibly inserted, so the access types are 'R' and 'I'. Following this, the Booking is created and linked to the Customer, so the access types are 'I' and 'L+'. Following this same principle, the Vehicle Category is read before being linked to the Booking, the 'to' and 'from' Local Offices are linked to the Booking and the Payment is created and linked to the Customer. The resulting Process Outline is shown in Fig. 7.19.

Rd Pth (read path)

This column and the next are read in conjunction with one another. The value n this column might be ambiguous without the qualifying value in the 'Access Via' Column.

The Read Path indicates the route by which this particular logical data group is accessed. The possible values for this are:

Dir   *Direct:* the logical data group is accessed directly by its prime key or data items specified in the Access Via column

Seq   *Sequential:* all the logical data group occurrences are read sequentially. This could also represent a retrieval of the logical data group sequentially following the previous logical data group on the Process Outline.

Det   *Via Detail:* the logical data group is accessed through its relationship with a detail logical data group. The detail logical data group is shown in the Access Via column.

Mast   *Via Master:* the logical data group is accessed through its relationship with a master logical data group. The master logical data group is shown in the Access Via column.

For the receipt of the booking, the majority of accesses are either via the customer or direct.

Access Via

If the Read Path is 'Via Master' or 'Via Detail', this column will contain the name of the master or detail logical data group that this logical data group was accessed from. If there is more than one relationship between the two data groups, the relationship used for the access is indicated here.

If the Read Path is 'Direct', this column will contain the data item(s) in the logical data group that are used as a key for the access.

The Process Outline for the receipt of the booking is shown with the Rd Pth and Access Via columns completed in Fig. 7.20.

No. Acc (number accessed)

This denotes the average number of logical data group occurrences accessed. The relative volumes on the Composite Logical Data Design will help to determine this number. If one occurrence of a master logical data group is accessed, the number of occurrences of the detail logical data group accessed will be the relative volume indicated on the relationship on the Logical Design Volume form (see Sec. 6.3). If accessing a master from a detail, the same number of master occurrences will be accessed as the detail as the one-to-many relationship between them means that each detail has only one master. The Process Outline for Receipt of Booking Request has only direct accesses, in which case only one will be read (assuming access is by prime key) or detail to master accesses which, again, will give rise to only one occurrence being accessed.

Data Items

The required data items within the logical data group being accessed are indicated here. If all of the data items are required, 'all' is specified here.

Logical Update Process Outline

System:	Date:	Author:	Event: Receipt of Booking Request

Access Type: (Acc Typ) — I - Insert, D - Delete, M - Modify, L+ - add Link, R - Read, L- - remove Link
Batch / On-line
Read Path: (Rd Pth) — Dir - Direct, Seq - Sequential, Det - via Detail, Mast - via Master

Op No.	Operation Description	I/O Ref	Acc No.	Entity	State Ind Val Pr	State Ind Set	Acc Typ	Rd Pth	Access Via	No. Acc	Data Items
O1 O2 O3	Check Customer exists. Display Credit Limit and accept confirmation to proceed. If 'off-the-street', create Customer	a-1	1	Customer	–	1	R I	Dir	Customer No.		
O4	Create Booking for Customer		2	Booking	–	1	I, L+	Mast	Customer		
O5	Check if requested Vehicle Category exists. If yes, link Booking to Vehicle Cat	E1	3	Vehicle Category			R L+	Dir	Vehicle Cat No.		
O6	Link to appropriate Local Office, using 'From' relationship		4	Local Office			L+	Dir	Office No.		
O7	If one-way hire, link to other Local Office using 'To' relationship		5	Local Office			L+	Dir	Office No.		
O8	If a deposit made, create Payment		6	Payment	–	1	I L+	Mast	Customer		

Fig. 7.20

Logical Update Process Outline

System:	Date:	Author:	Event: Receipt of Booking Request

Access Type: (Acc Typ) — I - Insert, D - Delete, M - Modify, L+ - add Link, R - Read, L- - remove Link
Batch / On-line
Read Path: (Rd Pth) — Dir - Direct, Seq - Sequential, Det - via Detail, Mast - via Master

Op No.	Operation Description	I/O Ref	Acc No.	Entity	State Ind Val Pr	State Ind Set	Acc Typ	Rd Pth	Access Via	No. Acc	Data Items
O1 O2 O3	Check Customer exists. Display Credit Limit and accept confirmation to proceed. If 'off-the-street', create Customer	a-1	1	Customer	–	1	R I	Dir	Customer No.	1	Customer No. Customer Name Credit Limit
O4	Create Booking for Customer		2	Booking	–	1	I, L+	Mast	Customer	1	Booking No. Date Received Driver Req/Not
O5	Check if requested Vehicle Category exists. If yes, link Booking to Vehicle Cat	E1	3	Vehicle Category			R L+	Dir	Vehicle Cat No.	1	Vehicle Cat No. Veh Cat Name
O6	Link to appropriate Local Office, using 'From' relationship		4	Local Office			L+	Dir	Office No.	1	Office No. Office Name
O7	If one-way hire, link to other Local Office using 'To' relationship		5	Local Office			L+	Dir	Office No.	1	Office No. Office Name
O8	If a deposit made, create Payment		6	Payment	–	1	I L+	Mast	Customer	1	All Items

Fig. 7.21

The completed Process Outline for the event Receipt of Booking Request is shown in Fig. 7.21. This represents a very detailed specification of the processing required for this event. In practice, it may not be necessary to fill all these columns in—the accessing and volumetric information is used in physical design to estimate the time taken to complete the processing. Only the more critical (complex, high-volume) Process Outlines are used for this, so it may not be necessary to complete all of this for the more simple, non-critical events.

Example of a Logical Enquiry Process Outline

Enquiry Process Outlines are produced in a similar way to the update Process Outlines. As can be seen in Fig 7.22 the form used is substantially the same as that used for the update. The only differences are the absences of the Acc Typ and the Set To State Ind columns. This is because enquiries will only use the Acc Typ, Read, and because changes to state indicator values are only caused by updates.

The enquiry Process Outlines are developed from the Retrievals Catalogue (see pages 125–6) which was developed in stage 2. Other documents that will be used are any relevant I/O Descriptions (the departure and return list was used as a data source for normalization in Fig. 6.15), the Composite Logical Data Design and supporting information, and the Logical Dialogue Outline (only developed for complex on-line enquiries).

Logical Enquiry Process Outline

System: Yorkies	Date:	Author:	Enquiry Name: Daily Departure and Return List

Batch / Online

Read Path: Dir - Direct
Seq - Sequential
Det - via Detail
Mast - via Master

Op No.	Operation Description	I/O Ref	Acc No.	Entity	Val Prev State Ind	Rd Pth	Access Via	No. Acc	Data Items
	For all Depots For each Depot	3.2-g		Local Office / Depot		Seq		50	Office No. Office Name Depot address

	Find all Bookings where Date Booking Start= Today (a) or Date Booking Ends=Today (b).
	If (a) write data items to Departures
	If (b) write data items to Returns
				Booking		Mast	Local Office / Depot from both 'To' And 'From' Relationships	50x
								x10=0
									Booking No.
									and more given on I/O Description

	As each Booking found: If Customer No. not null find Customer Name			Customer		Det	Booking	50x8	Customer Name
	If Driver No. not null find Driver Name			Driver		Det	Booking	5x6	Driver Name

Fig. 7.22

This example illustrates some of the difficulties involved in producing implementation-independent Process Outlines. In this case the most efficient access path for the retrieval to follow would be determined by the eventual data structure and the data management software—thus the supposedly logical Process Outline could be written in other ways (e.g. all Bookings could be searched serially for matchings with the required date). If a relational database is to be used it is probably just as easy to write enquiries like this one directly into SQL (see Sec.8.3).

Another question that arises is how to deal with conditional statements. Here there are iteration statements ('For all Depots') and selection statements ('If Cust No not null find Cust Name'). If Structured English is used then indentations are used to indicate the structure. Diagrammatic representations of Process Outlines are discussed below, and will almost certainly become a part of Version 4 of SSADM.

Note that in this example we have completed the No. Acc column. These figures are derived from those used on the Logical Design Volumes form (see Fig. 6.35) in the way described above. Obviously the actual number of occurrences accessed will depend on the eventual database structure, the software used, and the way the application program is written. However, these figures serve as a guide when converting the Process Outline to a physical design and help in predicting the performance of the program during physical design control.

Avoiding complexity on Process Outlines

The choice of access path is fairly clear for simple processes that access only a few entities, but for a slightly more complex process, the access path may not be so obvious. It may be necessary to jump in at different places in the Composite Logical Data Design, or there may be optional paths depending upon certain circumstances. If it becomes too complicated, the Process Outline will be impossible to read—and even more impossible to implement! One way to avoid this is to separate out the 'normal' processing from the validation and error-handling processing. The 'normal' path is reflected on a Process Outline and references are made (using the I/O Ref column) to the error processing detailed on another Process Outline.

Alternatively, a flowchart or Jackson-style structure expressing the access sequences might be useful as an overview of a complicated Process Outline. However, if a structured programming technique is to be used subsequently, care must be taken not to clash or preclude the use of such a technique. The more detailed the specification of processing sequence in a Process Outline, the more constrained a programmer will be in the design of the process logic.

Alternative ways of structuring the Process Outlines

The Process Outlines are structured around the accesses to the Composite Logical Data Design. As shown above, this access path can be indicated by a set of arrows annotating the Composite Logical Data Design. However, in the example above, the accessing structure was derived through no more than intuition, experience, and common sense. A complex Process Outline may require a very complex access path. When derived in this intuitive way, the Process Outline will almost certainly be unreadable and unstructured. A programmer trying to understand such a Process Outline to derive a satisfactory program structure might be forgiven for coding this unstructured logic directly in preference to unravelling the complexities of the Process Outline. Obviously, a preferable approach is to structure the Process Outline before the detailed operations are added. This might be approached in several ways including:

• the Process Outline form is replaced with a diagram supported by narrative;

- the structure of the Process Outline is derived first and then the form is completed as normal.

The Process Outline as a diagram This approach has been adopted within LSDM (LBMS System Development Method). The dialogue structures developed earlier are expanded to show small modules and their dependencies.

The structure derived before the Process Outline completed If Jackson Structured Programming is to be used to design the programs, this is the best technique to use to structure the Process Outlines. A suggested approach is described here.

- Decide upon the logical data groups to be accessed.
- Construct a Jackson structure to represent the sequence of accesses for the 'normal' transaction.
- Add into the diagram the accesses for error handling.
- Convert the diagram into structured logic or an action diagram.
- Transcribe this onto a Process Outline form and add in the extra information about the operations and accesses.

Use of Process Outlines with fourth-generation technology

The guidelines given above assume that it is appropriate for the analyst/designer to determine the access path for each process. However, there are an increasing number of development tools becoming available for which it is less appropriate to define the access paths. In this case, it is recommended that the Process Outline form is used without attempting to add any type of structure except that necessary from the Logical Dialogue Outlines.

The read path will not be needed in this case but it might still be useful to define the number accessed for timing calculations and the data items used to complete the specification.

SUMMARY

- Process Outlines are a complete logical specification of the processing required in the system.
- Logical Update Process Outlines are completed for each event identified on Entity Life Histories.
- Logical Enquiry Process Outlines are completed for all enquiries or retrievals.
- Process Outlines are structured around the accessing of logical data groups from the Composite Logical Data Design.
- An access path may be traced on a copy of the Composite Logical Data Design for each Process Outline.
- It is important to avoid complexity on Process Outlines so that they can be understood later.
- It is possible to structure Process Outlines in several different ways.
- The approach may be simplified if fourth-generation technology is to be used for the implementation.

EXERCISE

Scapegoat Systems

The Event Catalogue contains this description of the event: Allocation of Staff to Project. Entities affected: Allocation (Insert).

Employees are allocated to projects after a project manager has been appointed. The project manager is given access to the system to find employees with the appropriate qualifications.

Once identified the employee's current workload against projects is checked to see if they are available. When the experience and availability of the employee has been checked the project manager is able to select a project to allocate him or her to. The system will check that the person doing the allocating has actually been allocated as project manager for this project before proceeding. The Allocation can then be created and linked to Project and Employee.

Use the Composite Logical Data Design developed in Sec. 6.3.

1. Indicate the accesses required on the Composite Logical Data Design, annotating the accesses to show the sequence.

2. Complete a Logical Update Process Outline for this event showing clearly the sequence of operations and each entity that is read as part of this process. (Ignore any error handling that may be required.)

8. Physical design (stage 6)

8.1 Introduction

The review of the detailed logical data and process design at the end of stages 4 and 5 marked the end of the logical stages of SSADM. The method now moves from this logical design to a physical design. The logical design could be implemented on any hardware or suitable software, or in other words the logical design is implementation independent. In stage 6 this implementation-independent design is converted into a design specific for the particular hardware and software configuration selected. This implementation-dependent design is known as the *physical* design.

In the first five stages of a SSADM project considerable effort went into ensuring that the requirements of the user for functionality and information content were correctly determined, and into designing a system that would meet these agreed requirements. These requirements are carried through to stage 6. There are two main objectives to this stage:

- To produce a tuned design .
- To ensure that all the system components and plans are ready to enable system construction and implementation to be completed.

> **Warning** In stage 6 SSADM can get very technical. To be able to understand fully this chapter you will need a good understanding of at least one database management system, an understanding of the general principles involved in database systems, and some knowledge of software engineering. These topics are covered by many books and some recommendations are given in the bibliography.

Producing a tuned design

There is an analogy here with the design of car engines. We may have a requirement to build a sports car that has a top speed of 150 m.p.h. and an acceleration of 0–60 m.p.h. in less than 8 seconds. There will be other considerations that compete with these objectives such as the cost of building the car and the desired engine capacity. To produce a car that meets all these objectives the designers will go through several stages. They will design the engine on paper (nowadays probably using computer-aided design), they will model its performance, and will adjust the design until the modelled performance meets their objectives. Only then will they build a prototype engine, and test and fine tune its performance.

A similar approach is followed in systems development. In SSADM we develop a model of the physical system and then measure the performance of that model against objectives set by the users. When we have a model that performs in a satisfactory way then we are ready to go ahead and build the system, and test and fine tune it. The same reasoning underlies the thinking of the system developer and the engineer: developing and tuning of a model is cheaper and less risky than building several prototypes, and testing and tuning each one.

Operating on a model also has the advantage that a much greater range of possibilities can be tested.

In order to develop the model of the physical system the logical design is first converted to a physical one. This is done in a prescriptive way without regard for system performance. Thus step 610, *create first cut physical data design,* turns the logical data design into one specific for the particular database or file management package. Similarly step 620, *create program specifications for major transactions,* takes the logical process design for certain critical system areas and develops it into a design specific to the software chosen.

Steps 610 and 620 turn the logical design into an implementation-dependent design which should meet all of the users' requirements apart from those concerning performance. None of the users' functional and information requirements are lost by this transformation into a physical design. Figure 8.1 shows the steps of stage 6.

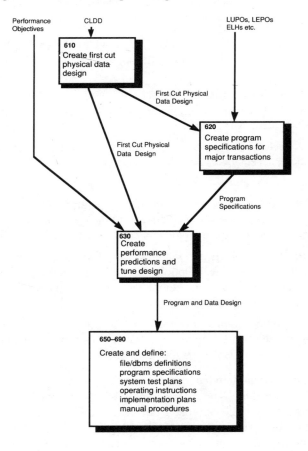

Fig. 8.1

In step 630, *create performance predictions and tune design*, the performance of that design is measured against performance objectives set by the users. The physical design may not meet these objectives: certain on-line responses may take longer than the 10 second maximum agreed with the users or the system may take up more than the 400 Mbytes of disk space agreed with the computer system manager. If the design does not meet these objectives then it is modified, or tuned, until it does. Thus this tuning is an iterative process in which several cycles of design, measurement, and redesign may be repeated before a satisfactory conclusion is reached. It may be that the objectives cannot be met with the planned hardware and software; in this case modification of the objectives will need to be discussed with the users.

Planning for systems development

When a design with acceptable performance has been produced, the full set of design documentation and plans needed for system construction and implementation can be developed. Thus in steps 640–690 the main activities are as follows.

Preparing for the construction

Step 640: create file/database definitions Once the design has been tuned to acceptable performance the file or database definitions can be loaded into the data dictionary or system library.

Step 650: complete program specifications Only partial Program Specifications for critical programs were developed before tuning. Full Program Specifications are developed for all programs in the new system.

Step 660: create systems test plan Detailed test plans are developed for integration, system requirement, acceptance, and volume testing.

Preparing for the changeover to the new system

Step 670: create operating instructions Operating schedules are created for each batch processing cycle in the system. These, and the operators' instructions for on-line programs, will form the basis of the operations manual.

Step 680: create implementation plans Plans for the changeover will have been considered in the Selection of Technical Options (stage 3) and possibly earlier in the project. However, once the test plans have been completed, detailed data conversion, changeover, installation, and other resource plans can be developed.

Step 690: define manual procedures The manual procedures that interface with the automated system will need to be designed. This step will also involve the development of a user manual and of plans for training the users.

Stage 6 and modern software environments

The way physical design is carried out will depend greatly on the kind of software used for the construction. The approach outlined above is particularly appropriate for a third-generation development in COBOL using a hierarchical database, network database, or conventional files. Figure 8.2 shows different development environments and their impact.

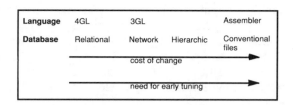

Fig. 8.2

Thus if assembler and conventional files are used, the cost of making changes to improve the performance will normally be much higher than if very high-level languages and a relational database are used. It also costs much more to make changes later in the development cycle (i.e. a change in the live system costs more than a change in the construction stage which costs more than a change in the design). Thus the more primitive the technology being used for construction, the more important it is to tune the system before construction begins.

If, however, the software used is so sophisticated that changes are easy to implement then the cost of early tuning may outweigh any benefits achieved. It then makes sense to develop a 'first cut' design into working code and tune the system throughout the construction stage.

Another approach is to follow the SSADM approach but to construct a prototype database and develop code for critical transactions. This prototype can then be tuned and can form the basis for the complete construction of the system.

In this chapter we follow the basic SSADM approach of tuning the system before detailed physical design and construction begin. However, this is an area in which software technology is changing very fast and therefore changes in stage 6 will occur in future versions of SSADM.

8.2 Create first cut physical data design (step 610)

The first step of stage 6 involves turning the Composite Logical Data Design into an executable file or database design. This is achieved by the application of simple rules, particular to the database management software (DBMS) or file handler used. The initial executable design is known as the 'first cut' because it is a first attempt at the physical data design. There may be several later versions of this design as it is tuned for optimum performance in step 630.

This first cut design is usually represented by a diagram supported by the amended Composite Logical Data Design Data Group descriptions. There are two basic criteria that the first cut design should meet:

1. It can be directly transformed to the data definition constructs of the software to be used.
2. It should contain sufficient information about the physical placement of data to enable performance calculations to be made.

The first criterion is concerned with being able to create a working system. All data dictionary systems, all database management systems, most programming languages, and most operating systems have constructs that enable data structures to be defined. The software then manages the placement and retrieval of that data on disc (or other physical storage media). These data definition constructs may be part of the programming language (as in COBOL), or may be a separate language as is the case with the CODASYL standard for network databases. Thus the first cut design diagram and supporting information can be transformed into the data definition syntax. This route is shown in Fig. 8.3. Some computer support tools to SSADM (and other methods) can automatically produce data definition statements from the logical data design held on the tool. This route is shown by the dotted lines on Fig. 8.3 below. Note that data definition statements can only be produced where data item lengths and format are known.

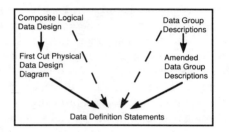

Fig. 8.3

In SSADM we normally develop a diagram as an intermediate stage before producing the statements because it is easier to apply the tuning to a diagrammatic representation. Monitoring the impact of various changes to the diagram is easier than monitoring the impact of changes to data definition statements. When the database design has been finalized the data definition statements are produced in step 640, *create file/database definitions*.

The second criterion is concerned with calculating the performance of the system. The time taken to perform certain critical transactions and batch programs needs to be calculated. Also the total space taken up by the system on backing storage needs to be calculated. To calculate both space and timings we need to know how the data is physically stored on disc.

Database management systems and file handlers usually store and retrieve data in the form of fixed size *pages* or *blocks* on disk. Similar groups of data (some software might call these *records*) are then stored on the same page. Often several data group occurrences can be stored on a single page. Sometimes it is possible to cluster data groups of different types together on the same page. The way these pages are organized often has a great impact on the performance of the system. Thus if the Customer data group occurrence and the related Order occurrences are stored together on the same page then all the information associated with a particular customer and his orders can be retrieved in a single disk access.

The first cut data design needs to show which data groups are clustered together so that data groups can be allocated to pages. Using the volumetric information collected in stage 4, the total number of pages, and thus the disk space required, can be calculated. All of these calculations are made in step 630, *create performance predictions and tune design*

First cut rules

Simple rules applied to the Composite Logical Data Design turn it into the First Cut Physical Data Design. The first cut rules will be specific to a particular software product. However, the rules are very similar within classes of software products. Thus all relational database management systems will have similar first cut rules. Below we give some classes of database management software and some general principles which the first cut rules follow. The hospital example used in Chapter 6 will be used to illustrate the basic principles. The dependent volumes are shown in Fig. 8.4 as these help decide how to cluster data groups.

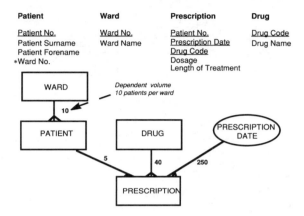

Fig. 8.4

Please note that the whole area of data management is riddled with technical jargon. DBMS vendors often use different words to mean the same things and the same words to mean different things. We have tried to be consistent and generally use the SSADM terms when describing the classes of product rather than the technical term. When technical terms are used they are explained (we have shown the reserved words of the DBMS products described in upper-case letters).

Hierarchical databases

The best-known example is IBM's IMS/VS which, at the time of writing, probably holds more gigabytes of data than any other DBMS. The database that underlies the fourth-generation environment of FOCUS is also hierarchical (although the FOCUS language can also interact with many other DBMSs).

These are the most complicated to convert to the physical design. This is because hierarchical databases only allow a data group to have one physical master. Thus in the hospital example above Prescription could either be physically owned by Patient or by Drug. The general principle when there are several possible masters is to pick the one for which there are fewest occurrences of the detail as the physical master. Patient is therefore picked as the physical owner of Prescription and Drug becomes the logical master. (Parent and child are the terms used by IMS to describe master and detail.)

In this way the data structure is divided into a number of physical segments. The data group at the top of the segment is known as the root segment. The linking across segments is provided by the use of logical pointers and secondary indexes. However, the rules for the use of these are complex, with several obvious ways of linking data groups being illegal. Figure 8.5 shows the first version of the IMS model of the hospital database; however, this is not yet an executable model as a logical child (Prescription) is not allowed to have a secondary index. To 'cure' this a 'Link' segment is created as physical child to Prescription whose logical parent is Drug. Further details of the precise rules for mapping logical data designs onto hierarchical databases will be found in the database books listed in the bibliography.

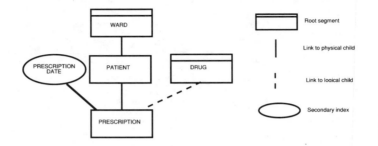

Fig. 8.5

Network databases

The most popular network database management systems are those that conform to the CODASYL standard set by the Database Task Group in 1971. The two most popular network database products in the UK are Cullinet's IDMS and ICL's IDMS, both of which follow the CODASYL standard.

The rules for converting the Composite Logical Data Design into a CODASYL design are very simple although there are a number of options concerning placement of RECORDs.

The basic rules are as follows:

- All data groups and operational masters become RECORDs.
- All relationships become SETs.

The remaining rules are concerned with placement of RECORDs. These can be placed in one of two ways:

CALC The RECORD is placed on the disk by a hashing algorithm which is applied to the key of the RECORD to give a physical disk address at which the RECORD is stored. The RECORD is retrieved by using the same algorithm.

VIA The RECORD is placed on the disk near its master in the specified SET.

All RECORDs that are at the top of the data structure (have no masters) are placed by the CALC mechanism. Also all RECORDs on which direct access is required using the key data item (sometimes called access entry points) are placed CALC. These are usually the RECORDs which have single item keys or composite keys. CALC RECORDs are indicated in the diagram by a stripe across the top of the box. This is the notation SSADM uses; it is related to but not the same as notations used by the vendors of CODASYL databases.

In the hospital example Ward, Drug, and Prescription Date are all at the top of the structure and are access entry points. Patient is an access entry point. All these RECORDs are placed by the CALC mechanism.

All other RECORDs must be placed VIA a SET. Thus the Prescription RECORD must be placed VIA a SET. Here there are three possible SETs that Prescription could be placed VIA. It can only be placed VIA one of these, in this example Prescription occurrences can only be placed close to their master Patient, or to their master Drug, or to their master Prescription Date. The rule for first cut design is to place the detail RECORD close to the master RECORD for which there are least occurrences of the detail. Referring back to Fig. 8.4, which gives the relative dependencies, shows that there are 5 prescriptions per patient, 40 prescriptions per drug, and 250 prescriptions on a given prescription date. Thus Prescription is placed physically close to Patient, VIA the SET linking them together. The SSADM notation is to show the SET the RECORD is placed VIA by a continuous arrow. Other SETs, which do not indicate placement, are shown by dotted arrows. Figure 8.6 shows the CODASYL representation of the hospital database.

Recent versions of these CODASYL databases include a secondary indexing facility. This is particularly useful in that it means that operational masters can be implemented directly rather than by creating a separate CALC RECORD. Thus in the hospital database we would create an index on Prescription of Prescription Date. The secondary index would be shown as an oval, as in the hierarchical data base example above.

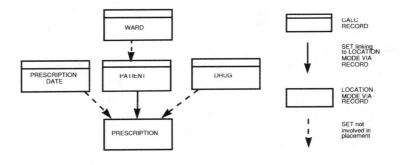

Fig. 8.6

Relational databases

In Chapter 6 we discussed relational data analysis and began Sec. 6.2 by explaining some aspects of the relational model. This model is that used by relational database management products such as ORACLE, INGRES, DB2, and many more. Relational databases are now widely accepted as the preferred approach with most projects using SSADM being implemented using a relational DBMS. All data is represented and manipulated as tables with relationships between tables of different types being managed by matching the values of data items in one table with those in another. A single language, usually called SQL, is used for data definition and data manipulation.

Although there is an ANSI standard SQL, there is considerable theoretical debate over the imitations of the standard, with some experts claiming the standard to be a 'lowest common denominator'. However, although there are many implementations of SQL they follow the standard sufficiently for the first cut rules to apply to most relational DBMSs.

Where relational DBMS products show greatest variation is in the underlying mechanisms used for storage and retrieval of data. Although these are hidden from the programmer or end-user who interacts with the system through SQL, they have a great impact on the performance of the system. Products vary greatly in the control they allow to the database administrator over such things as storage structures (INGRES allows tables to be organized as B*-trees, hash files, indexed sequential files, sequential files, or as heap files; ORACLE organizes all tables as B*-trees), indexing, page size, and clustering of records. With some products, particularly those from third-party vendors such as ORACLE and INGRES which run under a large range of operating systems, some of these storage issues will be dependent on the operating system rather than the database management system. The first cut rules for determining the underlying storage will therefore be different for each product and operating system combination. These may, of course, be subject to change with later software releases.

Below we give the first cut rules for the ORACLE relational database management system and apply these to the hospital database. We give both a diagrammatic representation of the the database structure (Fig. 8.7) and some sample SQL PLUS (Fig. 8.8). This is the language used by ORACLE for data definition and manipulation and is an extension of ANSI SQL. Thus the rules and solutions given will be fairly similar to those for other relational products.

In stage 4 of SSADM, detailed data design, a well-normalized data structure was produced and merged with the Logical Data Structure to form the Composite Logical Data Design. This will be already in a form very close to that required by relational products. The transformation of the Composite Logical Data Design to a relational first cut design is therefore very simple.

The basic rules are:

1. Ensure that all primary and compound keys are present and defined for each data group.
2. Each data group (except for operational masters) is defined as a TABLE.
3. All items contained in a primary key are defined as NOT NULL. Foreign key items are defined as NOT NULL unless they indicate an optional relationship (they could then take a NULL value).
4. Create a UNIQUE INDEX for each primary key (if it is a compound key the INDEX must be over each data item within it). Create an INDEX for each foreign key.

To illustrate how straightforward it is to develop the data definition statements for relational databases we give the SQL PLUS statements for this database in Fig. 9.10. We have used the default options for INDEXes; these would probably be changed in the tuning step. We have allowed the non-key data items to default to allowing NULL values; it would be sensible to enforce values if this was required.

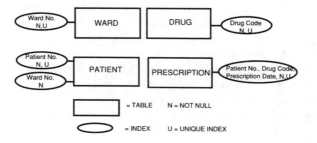

Fig. 8.7

```
CREATE TABLE Patient (Pat_No NUMBER (5) NOT NULL, Pat_Surname CHAR (20),
Pat_Forename CHAR (20), Ward_No NUMBER (5) NOT NULL)
CREATE TABLE Ward (Ward_No NUMBER (5) NOT NULL, Ward_Name CHAR (20))
CREATE TABLE Drug (Drug_Code CHAR (5) NOT NULL, Drug_Name CHAR (30))
CREATE TABLE Prescription ( Pat_No NUMBER (5) NOT NULL, Drug_Code CHAR (5) NOT
NULL, Prescription_Date DATE NOT NULL, Dosage CHAR (20), Len_Trtmnt NUMBER (3))
CREATE UNIQUE INDEX Patient_PK ON Patient (Pat_No)
CREATE UNIQUE INDEX Ward_PK ON Ward (Ward_No)
CREATE UNIQUE INDEX Drug_PK ON Drug (Drug_Code)
CREATE UNIQUE INDEX Prescription_PK ON Prescription ( Pat_No , Drug_Code ,
Prescription_Date)
CREATE INDEX Patient_FK ON Ward (Pat_No)
```

Fig. 8.8

SUMMARY
- A First Cut Physical Data Design is produced by applying simple rules to the Composite Logical Data Design.
- These first cut rules are specific to the DBMS or file handler used.
- No attempt is made to optimize performance with the first cut design.
- The first cut design is sufficiently detailed to allow performance predictions to be made and enable data definition statements to be produced.

8.3   Create Program Specifications for major transactions (step 620)

This involves identifying the most critical processing in the system and converting the logical specifications of that processing to physical specifications. These can then be used, in step 630, to create detailed predictions of the time taken to perform these transactions by the chosen hardware and software. This step continues the program specification started at an implementation independent level in stage 5, with final Program Specifications being produced after tuning in step 650. Thus Program Specifications are produced in two phases:

Before tuning—programs for critical transactions are specified in sufficient detail for timing

After tuning—previously specified programs may be modified and all programs are completely specified

In this section we describe the techniques that are used throughout program specification in SSADM, these are used both before and after tuning.

Selecting the major transactions

There is no point in developing Program Specifications for all of the processing at this time. As a result of tuning, the database structure may well be changed to improve performance and this may have an impact on the processing. For instance, in tuning, we may decide to include redundant links between data groups or to hold derived data to improve the response speed for certain transactions. Other transactions will affected by this, they will have to maintain the derived data and the links or they may be able to make use of the change in structure. So if all the processing was specified at a detailed physical level at this time, it would certainly need drastic revision after tuning.

In many systems there are a few programs which are executed many times, occupying perhaps 80 per cent of the computer's resources but only comprising perhaps 20 per cent of the actual program code. The performance of these programs is therefore critical to the performance of the system as a whole. It is important to identify these critical programs and particularly the most frequent transactions within them. These transactions are the ones which will be specified at a physical level so that their performance can be measured and improved.

Other transactions may be critical, not because of their frequency but because a fast response is required—perhaps for rapid decision making or as a 'real time' component of the system.

Thus the basic criteria for selecting the transactions which need to be specified in detail and then tuned are:

- high volume on-line transactions;
- major daily batch runs;
- programs run at peak times;
- on-line transactions requiring a fast response or real-time transactions.

To decide what processing falls into the above categories we use some of the volumetric information collected in the earlier stages of the project. Some sizing of the processing load will have been performed in the technical options stage (see page 170) and these calculations should be reviewed as part of the selection of critical transactions. Performance Objectives were also developed as part of stage 3 and these, to some extent, predicted the critical transactions of the system. As well as deciding which were the high volume transactions, timings for on-line transactions requiring a fast response are agreed with the users.

At the end of stage 2 an Event Catalogue was developed (see pages 148–9). This included maximum and average frequencies for each event. These volumes may have been refined during the technical options and detailed process design stages. This information gives some idea of the volumes of particular update transactions (we can think of an update transaction as the eventual physical implementation of an event). However, we need to go elsewhere to get volumetric information about the enquiry transactions and about how each transaction will become a part of an on-line or batch program.

Volumetric information about the enquiry transactions (or retrievals) is often hard to predict, particularly with management information or decision support systems, since users may use the system in a very different way to that anticipated. Some indications may come from the way the current system is used and from the users' estimates of enquiry frequency. This information will have been associated with the Retrieval Catalogue (see pages 125–6) or with the Logical Enquiry Process Outline developed in step 510.

The way transactions are grouped together to form batch or on-line programs is defined by the Function Catalogue. Using this it is possible to tell whether transactions are to be performed on-line or in batch mode, and if batch which processing cycle they belong in. This is discussed further in the next section.

Using the above approach we can determine which transactions are the most frequent and when they are going to be run. Thus the high volume on-line transactions, the major daily batch runs, and the programs run at peak times can be found.

Logical processes to physical programs

Before physical design, all the processing has been defined independently of the hardware/software environment. This logical processing has been documented by Logical Update and Enquiry Process Outlines, Logical Dialogue Outlines and Controls, Data Flow Diagrams, and by Function and Retrieval Catalogues. These need to be converted to the selected physical environment. As was discussed above this is done in two stages, initially for the critical processing and after tuning for all the processing. In this step we are concerned with converting primarily the functions and the Process Outlines.

Functions become programs

As part of the specification of the required system processing (step 230, Sec. 4.5), Data Flow Diagram processes were allocated to high-level functions. These are groupings of low-level processes that may be physically implemented together. As such they may be thought of as forming the programs of the new system. The extent to which the eventual programs, as they are coded, will follow this structure will depend on a number of factors such as: the programming language, DBMS, operating system, the structure, experience, and size of the programming team, and installation standards. However, this first attempt at deciding what goes into each program is a necessary step towards the eventual program design.

Each function is made up of a number of low-level processes. The structure of the functions may not follow that of the processes. This is because within a high-level process there may be lower-level processes; some of which will be performed on-line, some in batch weekly mode, some in batch daily mode. Obviously these could not be combined into a single program. There may be other reasons why the particular Data Flow Diagram process hierarchy does not lend itself to the program structure.

However, in most simple systems, which are mainly on-line, the function structure and therefore the program structure follows the structure of the processes. This is true of the Yorkies case study where the only difference between the process structure and the function structure is in process 4, Obtain Money, where the sub-processes 4.2 and 4.4 concerned with producing invoices and reminders are performed in batch mode and the other sub-processes concerned with recording payments are performed on-line. The Yorkies Function Catalogues are shown in Fig. 4.15 and described on pages 124–6.

Because it is the Process Outlines, not the Data Flow Diagram processes, that form the basis of program specification in SSADM we need to establish which function each Process Outline belongs to. This can easily be done by inspection of the Event Catalogue and the Function Catalogue. The Event Catalogue shows which processes handle which events and the Function Catalogue shows which processes belong in each function. As every event has an Update Process Outline it is a simple matter to allocate the Process Outlines to Functions. Any problems can be resolved by reference to the Data Flow Diagrams.

Each program identified is described at a high level by a Physical Program Description. This is usually one page of A4 narrative describing what the program does, when it is used, who uses it, and any other general information about the program. This documentation will be modified and extended as the specification is developed in physical design.

Logical Process Outlines become Physical Process Specifications
The detailed process design (stage 5) described the system processing using Logical Process Outlines. In physical design we extend these to become implementation-specific designs known as Physical Process Specifications. Basically this involves modifying the Process Outline form completed in stage 5 to include physical design-related information.

There is usually no need to transcribe the information to a new form as the only differences between the Update Process Outline form and the Physical Process Specification form are in the column headings. Acc Typ in the Process Outline is replaced by Process Verb to indicate the verb used by the DBMS or programming language to perform the access (so READ would be replaced by the CODASYL verb, OBTAIN, or the SQL verb, SELECT). The ENTITY heading might also be replaced by the name used by the DBMS (so RECORD TYPE might be used by CODASYL and TABLE by SQL).

Conversion of a Process Outline to a Physical Process Specification usually requires the addition or modification of the following physical design-related information.

Retrieval logic How the required information is retrieved using the particular first cut design for the DBMS chosen. This may only involve specifying the mechanism to be used in the Read Path column.

Storage logic For updates the process verb will always need to be specified. Sometimes it will be necessary to specify how links are made between different occurrences. It is often best to avoid writing any information to disk until the end of a transaction; this facilitates recovery should a failure occur.

Manipulation logic and operation sequences The way the information is manipulated and the sequence of operations may need to be changed from the Process Outline because of the programming language selected or because of constraints imposed by the DBMS on the first cut data design.

Screen and menu handling For on-line processing the display of information and messages needs to be considered: if a Logical Dialogue Outline was produced then this may also need modification. The Physical Process Specification and the Logical Dialogue Outline complement each other and should be consistent.

Logical Enquiry Process Outline

System: Yorkies   Date: Author: Enquiry Name: Daily Departure and Return List

Batch / Online   Read Path: Dir - Direct / Seq - Sequential / Det - via Detail / Mast - via Master

Op No.	Operation Description	I/O Ref	Acc No.	Entity	Val Prev State Ind	Rd Pth	Access Via	No. Acc	Data Items
	For all Depots / For each Depot	3.2-g		Local Office / Depot		Seq		50	Office No. / Office Name / Depot address
	Find all Bookings where Date Booking Start= Today (a) or Date Booking Ends=Today (b). / If (a) write data items to Departures / If (b) write data items to Returns			Booking		Mast	Local Office / Depot from both 'To' And 'From' Relationships	50x2 x1050	Booking No. and more given on I/O Description
	As each Booking found: / If Customer No. not null find Customer Name			Customer		Det	Booking	50x8	Customer Name
	If Driver No. not null find Driver Name			Driver		Det	Booking	50x6	Driver Name

CODASYL Physical Process Specification

System: Yorkies   Date: Author: Event Name & Id: Daily Depature And Return List

Batch / Online   Process Verbs: F- FIND / G- GET / O- OBTAIN / M- MODIFY   C- CONNECT / D- DISCONNECT / E- ERASE / S- STORE   Read Path: C- CALC / N- NEXT pointer / F- PRIOR pointer / O- OWNER pointer

Op No.	Operation description	I/O Ref	Acc No.	RECORD TYPE	State Ind Val Pre	State Ind Set To	Proc Verb	Rd Pth	Access Via	No. Acc	Data items
	Follow system owned SET to find all Depots / As each Depot found write items to P-file	3.2-g		Loc Office/ Depot			O	N		50	Office No. / Office Name / Depot address
	Follow 'from' SET to find all Bookings where Date-Booking-Start= Today / Write data items to Departures in P-file / As each Booking found			Booking			O	N	Loc Off/ Depot using 'from' SET	50 x1050	Booking No. and more given on I/O Description
	FIND OWNER in Customer SET write Customer No., Customer Name to Departures			Customer			O	O	Booking	50x8	Cust No. / Cust Name
	FIND OWNER in Driver SET write Driver No., Driver Name to Departures			Driver			O	O	Booking	50x6	Driver No. / Driver Name
	Follow 'to' SET to find all Bookings where Date-Booking-Ends=Today / Write data items to Returns in P-file / As each Booking found			Booking			O	N	Loc Off/ Depot using 'to' SET	50 x1050	Booking No. and more given on I/O Description
	FIND OWNER in Customer SET write Customer No., Customer Name to Returns			Customer			O	O	Booking	50x8	Customer No. / Customer Name
	FIND OWNER in Driver SET write Driver No., Driver Name to Returns			Driver			O	O	Booking	50x6	Driver No. / Driver Name

Fig. 8.9

In Fig. 8.9 we have shown how the Enquiry Process Outline, 'Daily Departure and Return List', has been converted into a Physical Process Specification. Note that the first cut database design has led to changes in the logic of the processing; the 'from' and 'to' SETs are followed separately to find the departing and returning vehicles. The Process Verb column is only useful, for enquiries, if the DBMS has a variety of READ verbs. (CODASYL uses FIND to place a cursor on the RECORD and then GET brings the the RECORD into working storage, OBTAIN does a FIND followed by a GET. In this case we have needed to examine all the RECORDS found so OBTAIN has been used, represented by an 'O'.) The Read Path column shows the method used by the DBMS for navigation, in this case NEXT and OWNER pointers have been used.

If a relational database is used it is often easier to code the SQL directly rather than develop a Physical Process Specification (or even a Logical Update Process Outline). This obviously saves time since the SQL can be used in the final implementation but also enables the performance to be tested on a prototype database. This is probably more accurate than the traditional paper tuning.

We illustrate this below with an ORACLE SQL query to retrieve information required for the Daily Departure and Return Lists. This query could be embedded in a programming language (e.g. C) or a report generator (e.g. SQL*Report). The retrieved values would then be put into variables and printed using the programming language or report generator facilities. The departures can be separated from returns in a number of ways: e.g. Veh_Reg_Mark or Start_Mileage are null for departures and not null for returns.

```
SELECT Office-No, Office_Name, Depot_Address, sysdate,/* an ORACLE pseudo-column -
today's date */ Booking-No, Booking.Customer-No, Customer_Name, Veh_Cat_Req,
Start_Date, Finish_Date, Driver_Req, Veh_Reg_Mark, Agency_Name, Booking.Driver-No,
Driver_Name, Start_Mileage, Date_Time_Collected,/* date type includes time */ Office-
No_Start, Office-No_Finish
FROM Local_Office, Booking, Customer, Driver
WHERE          Booking.Customer-No = Customer.Customer-No(+)
AND   Booking.Driver-No = Driver.Driver-No(+)
AND   (                      (Start_Date = trunc(sysdate)
              AND       Office-No=Office-No_Start)
       OR                    (Finish_Date = trunc(sysdate)
              AND       Office-No=Office-No_Finish))
AND   Office-No = local_office_no /* value to be supplied */
```

Fig. 8.10

Note

1.   Where the relationship is optional (i.e. the foreign key can be null), an outer join ((+)) is used to ensure that the rows with null foreign keys are also selected.
2.   The pseudo-column sysdate has the time truncated to zero. This is because
    (a) time is taken into account when comparing date types;
    (b) when date types are entered without the time, the time is set by ORACLE to zero (e.g. Start_Date and Finish_Date).

Common processing

One problem that often occurs during systems development is the identification of common processing. Obviously it would be very wasteful to reinvent specifications and code that has been used elsewhere. This could also lead to inconsistent processing. The solution suggested in SSADM is to develop a Common Processing Catalogue in which specifications of any processing likely to recur are recorded. This may be started in the development of the Required System Data Flow Diagrams and supporting documentation but should certainly be used when Process Outlines are developed. The common process is referenced in each of the Process Outlines that use it.

On-line program specification

Each on-line function becomes an on-line program. Processes will have been grouped together by operating on similar data, by performing similar processing, and by being used by the same group of users. Each function will normally become, in a menu-based interface, a high-level option with each of the individual events becoming lower-level options. This kind of structure is shown below in Fig. 8.11.

```
Function 1 ──┬── Event A  + Physical Process Specifcation
             ├── Event B etc.
             └── Event C etc.

Function 2 ──┬── Enquiry A + Physical Process Specification
             ├── Enquiry B etc.
             └── Enquiry C etc.

Function 3 etc.
```

Fig. 8.11

Even if the interface will be based on commands rather than menus, thinking of the structure of events and functions in this way can be helpful. Logical Dialogue Controls, if they have been developed in stage 5, can also be used to show the possible sequences of transactions. These may need modification because of the system software (fourth-generation system, transaction processing monitor, etc) used. Some basic fourth-generation systems are only able to handle hierarchic menus with the user confined to returning to the top-level menu before moving to another option. More sophisticated software will allow direct navigation from one transaction to any other by use of commands, will allow passing of data between transactions, and will employ advanced human computer interfaces such as windows, pop-up-menus, touch screens, pointers, and icons.

In step 620 only the critical transactions need be specified in enough detail for them to be timed. This means converting the Process Outlines to Physical Process Specifications in the way described above. Screen and menu handling logic will need to be added to the Physical Process Specifications; the I/O Descriptions and any Logical Dialogue Outlines created in stages 2 and 5 will help. It is probably not worth converting I/O Descriptions to screen formats yet as these may need to be changed as a result of tuning.

Some large systems might use on-line creation of transaction files before database update. As these transaction files are not part of the main system data (they are not shown on the Composite Logical Data Design) they may not have been considered in stage 5. If so there will be no Logical Update Process Outline and the Physical Process Specification will need to be created from documentation associated with the Data Flow Diagram such as the I/O Descriptions, Elementary Function Descriptions, and the diagram itself. The content and organization of any transaction files produced will also need to be specified.

Batch program specification

Before tuning it is necessary to specify the critical batch runs. These will probably be all of the major processing cycles such as the daily, weekly, and end-of-month runs. In general it is more complex to specify the batch programs than the on-line programs. This is because there will usually be greater interaction between the various transactions which are handled by the program and a greater interaction between the programs, with the output from one becoming the input to another program. It is also necessary to define the various transaction files and specify how they will be used.

Conversion of Process Outlines to Physical Process Specifications follows the same guidelines as for on-line programs. However, there will usually be some transaction file processing which has no corresponding Process Outlines. Development of the Physical Process Specifications for these will usually be based on the documentation associated with the Required System Data Flow Diagrams.

The design of a each batch processing cycle is usually represented by a systems flow chart. This uses a wide range of symbols to show different kinds of storage media, different ways of processing data, and different kinds of input and output. It is a good and standard way of showing the typical activities in a batch cycle such as sorts, merges, and use of transaction files. Systems flowcharts differ from Data Flow Diagrams in that they explicitly show sequences and decisions.

The first step in batch program specification is to decide the basic sequence of programs within each cycle. System input and output processing and their associated transaction files can then be considered. In this way a run flowchart can be built up, in a step-wise way, for each processing cycle.

1. Define the basic run flows The suggested approach is to use the Entity Life Histories to help define the sequence. As these show the possible sequences of events that can affect each entity we can draw up maps of possible sequences of functions within each processing cycle. The easiest way of doing this is to mark the events on each Entity Life History with the batch function number and then use these to produce a set of maps for each processing cycle. The sequence maps for each cycle from all the Entity Life Histories can then be merged to give the basic run flows. The run flow for each cycle, represented as a systems flowchart, can then be checked against the required system Data Flow Diagrams.

2. Deal with input transaction files Firstly these need to be identified. The source data for each batch program will be either the database or transaction file(s) or both. Some transaction files may be shown on the Required System Data Flow Diagrams. We need to specify these files in terms of their data content, organization, and storage medium. Information associated with the Data Flow Diagrams such as the I/O Descriptions and the Elementary Function Descriptions will help create the file specification and the Physical Process Specification if no Process Outline has been developed.

Within each program we need to decide on the transaction sequence, i.e. in what order are the various events to be processed by the program. A frequent technique used in batch processing is to sort the input transaction file by entity occurrence. Thus if we were updating a Customer master file we would sort the input transactions by Customer No. and perform all of the processing for each Customer before passing onto the next one. The traditional order is then to perform the deletions first followed by the insertions and modifications. In this way, if Customer Nos. are being reused, there is no possibility of a new customer being created and deleted within the same processing cycle. The Physical Process Specifications may need to be structured in a different way to the Process Outlines if a batch program is to be performed this way. Any sorts should be added to the system flowchart as it is being built up.

3. Deal with program output   Output transaction files from a program may be input to other programs in the same system; they may be used by other systems inside or outside the organization; or they may be reports and log outputs going to spooling or printing systems. These outputs need to be specified in terms of their format and content, I/O Descriptions will help with this task. The Physical Process Specifications created in task 2 above will need to be amended and extended to take account of this output processing. The run flowchart will also be extended by showing the output processing and transaction files.

4. Transaction file maintenance   There may be other transaction file maintenance to be considered such as input edits, transaction file reformats, and extensions. No Process Outlines will have been developed for these so the Physical Process Specifications will be based on the Data Flow Diagrams and their supporting documentation. These programs are added to the run flow.

5. Batch enquiry programs   These are specified in a similar way to the update programs. The Retrievals Catalogue, Logical Enquiry Process Outlines, I/O Descriptions, and possibly the Data Flow Diagrams will help. Some batch programs may create and use transaction files, these and any resulting sorts should be added to the run flow of the appropriate processing cycle. An example of a Physical Process Specification for a batch program was shown in Fig. 8.9.

SUMMARY

- Program specifications are developed for the critical transactions.
- Critical transactions are ones that have a significant impact on the performance of the system, particularly high-volume, on-line transactions and major batch runs.
- Specifications are developed in sufficient detail for performance to be predicted so that tuning can be done.
  Functions developed in stage 2 become programs in stage 6.
- Process Outlines developed in stage 5 are converted to Physical Process Specifications in stage 6 by adding physical design-related information.
- On-line programs are specified with the help of Logical Dialogue Outlines and Controls.
- Batch programs are more complex to specify because each processing cycle and many transaction files have to be defined.
- The basic run flow is created using the sequences defined by the Entity Life Histories.
- The run flow is expanded to include the processing of input and output transaction files.
- Each batch program is further specified by a number of Physical Process Specifications

8.4   Create performance predictions and tune design (step 630)

Introduction

The objectives of this step are to produce a tuned physical data and process design that meets the Performance Objectives agreed with users in stage 3 and makes the best possible use of the hardware and software available.

The basic approach to achieving this is to measure the performance of the initial physical design for data and processing against the Performance Objectives. These designs can then be adjusted until the objectives are met. This process is known as physical design control (illustrated by the flowchart in Fig.8.12). If it is impossible to modify the design sufficiently to meet the Performance Objectives then changes to these will need to be discussed with users. This may mean a reduction in service levels or the purchase of additional hardware.

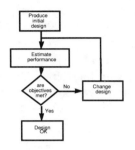

Fig. 8.12

There are four basic inputs to the step.

First Cut Physical Data Design Developed in step 610 as the first attempt at developing an executable data design. It is sized in terms of the storage space required. The time taken for critical transactions to be performed on it is estimated. This design will be modified if it does not, as is usually the case, meet the Performance Objectives.

Program Specifications for major transactions These were developed in step 620 in sufficient detail for timing to take place. They will normally need adjusting to be able to meet the Performance Objectives.

Performance Objectives These were developed in step 340 after the technical options had been decided. The users, assisted by the development team, defined objectives for data storage, timings for critical transactions, recovery, and information.

Hardware and software performance statistics These are needed to calculate the storage requirements and transaction timings. Installations should develop statistics for their own range of hardware and software, otherwise this information should be available from the supplier.

The output from the step will be: a modified First Cut Physical Data Design, modified Program Specifications, and possibly agreed changes to the Performance Objectives.

Measuring performance against objectives

The performance of the system is measured against the Performance Objectives set in stage 3 (these are described in Sec. 5.5). Initially this means estimating the performance of the First Cut Physical Data Design and initial Program Specifications. If the Performance Objectives are not met then these designs will be modified and estimated again. This iterative process of performance estimation and design modification is continued until either the Performance Objectives are met or it becomes apparent that no improvement is possible.

Some objectives that may be set, such as those for information or portability of software, are not directly measurable and are not considered here. Performance Objectives are also set for storage space, function timing, and recovery. These are quantifiable and we will describe how the performance of the system can be estimated against each of these in turn.

Estimation of storage requirements

Most projects will make some estimates of the space required as part of the project justification. These will usually be very rough estimates based on the number of occurrences of the major entities. As more information is collected and a more detailed view of the new system is developed, then more accurate sizing can be done. In stage 3 each of the technical options is sized (described on page 170) and the sizing of the chosen option is often used as the basis for the storage space performance criteria.

In this step the First Cut Physical Data Design is sized in a more detailed and accurate way. The sizing is based upon the allocation of data group occurrences to specific areas of disk storage.

Below we describe the information that is required for this detailed sizing. This basically comes from three sources: the Logical Design Volumes (see pages 207–10), the First Cut Physical Data Design, and from the hardware and software environment to be used.

From the Composite Logical Data Design and Logical Design Volumes detailed information about the data volumes are extracted:

- data space for each data group;
- volumes of each data group ;
- volumes of relationships;
- variance of volumes over time;
- which data needs to be on-line.

From the First Cut Physical Data Design we need to know:

- how data groups are clustered together;
- volumes of any new records introduced in first cut design.

Factors required from the hardware and software environment are:

- possible page sizes, buffer allocation;
- DBMS overheads for page management;
- DBMS overheads for indices, pointers, hashing ;
- handling of insertions, deletions, overflow, etc.;
- DBMS data type handling and data compaction possibilities.

The way that most database management systems, file handlers, and operating systems handle the retrieval and organisation of data on disk is through a set of fixed size *pages* (operating systems usually refer to them as *blocks)*. The page is the unit of disk I/O, with data written to buffer a page at a time to be manipulated and then written back to disk for update. The allocation of data groups to pages therefore has a great impact on the speed of retrieval as well as on the space required.

CODASYL page planning chart											
Page size: 8k		Page type: Customer			No. CALC RECORDs: 3000		No. Pages: 3000		Expansion by 40 % =	4200 pages	
RECORD	Data size	DBMS o'head	Pointers				Total pointer space	Total RECORD space	No. RECORDs per CALC	Space per CALC	Comment
			CALC	NEXT	PRIOR	OWN					
Customer	124	6	4	4			8	138	1	138	
Payment	43	6		8		4	12	61	27	1647	
Invoice	57	6		12		4	16	79	28	2212	
Payment / Invoice link	18	6		8		8	16	40	28	1120	

Total	5117	
CALCs / page	1	
Total data space	5117	+ 32
Page space	5149	

Fig. 8.13

Generally each system software product handles page management in a different way. Differences occur in a number of areas such as: possible page sizes, how each page is organized, and in the numbers of different data groups that can be stored on a single page. In order to plan the allocation of data groups to pages it is suggested that page planning charts are developed specific to the software. One of these charts will then be completed for each cluster. In Fig. 8.13 we have shown a completed CODASYL page planning chart for a portion of the Yorkies First Cut Physical Data Design. (Note to the technically minded: we have assumed 4 bytes for all pointers, only implemented NEXT and OWNER pointers in the first cut design, and each CALC record belongs to a system owned SET.)

A chart is compiled for each cluster in the First Cut Physical Data Design. As each one shows the total number of pages occupied by the cluster, adding all these up gives the total number of pages occupied by the database. Then multiplying by the page size gives the database size in bytes.

Because this chart is for CODASYL databases it shows space for pointers. Many relational databases (e.g. ORACLE, DB2) use indexes as their primary organization. For these there would be two types of page planning charts, one for the index pages and one for the data pages.

Estimation of function timings

Some rough estimates of function timing will have been done in stage 3. These may have been used to help define the Performance Objectives. In this step the timings for the critical functions are estimated in more detail.

Below we describe the information that is required for these detailed timings. This basically comes from four sources: the Physical Process Specifications and Program Specifications, the First Cut Physical Data Design, the Event Catalogue and enquiry volumes, and from the hardware and software environment to be used.

For timing batch runs we will need the run flowcharts, the specification of each program, and each of the Physical Process Specifications making up the program. For timing the critical on-line transactions usually only the Physical Process Specifications are required. The information below is extracted from the Physical Process Specification:

- access path through First Cut Physical Data Design;
- access and retrieval mechanisms;
- storage logic;
- manipulation logic;
- process verbs used;
- numbers of records accessed;
- screen handling.

The First Cut Physical Data Design and the page planning charts tell us which records are clustered together.

The Event Catalogue and the enquiry volumes give us the transaction volumes and the variance of those volumes over time.

Factors required from the hardware and software environment are:

- CPU time per DBMS call;
- CPU time for application program per DBMS call;
- CPU time per message pair (for on-line);
- disk access time;
- record locking mechanisms;
- security overhead time (for preventing unauthorized access, etc.);
- recovery overhead time (for writing to transaction logs, etc.);
- data communication line speeds.

These sort of figures can be quite hard to come by. If the installation is using the same hardware and software for another system and has a good performance monitor, then installation standards can be set for the above. If hardware and software are to be purchased, the above statistics may be available from the supplier or there may be benchmark figures available from which these statistics can be derived.

The basic approach to timing is to calculate the disk access time and CPU utilization time for each transaction. The actual average elapsed time taken by a transaction (or response time for an on-line transaction) will probably be several times larger than that calculated. The elapsed time will be affected by such factors as queuing time, record locking, the overall machine loading which may governed by other systems using the same machine, and by peaks and troughs. The prediction of overall system performance is a difficult area and there are several simulation programs and expert systems programs available which attempt to predict and improve system performance either generally or for specific hardware and software.

For critical on-line transactions the timings calculated can be compared with the objectives set for those transactions. For batch programs the times taken to perform each Physical Process Specification are aggregated to give a total for the program. All the programs within the cycle can then be aggregated to give a figure for the whole cycle. In Fig. 8.14 we illustrate a transaction timing form that can be used to estimate the performance of transactions on a CODASYL database. We have used the Daily Departure and Return List enquiry that we developed the Physical Process Specification for on pages 247–51. The figures given below for database performance and disk access are reasonable at the time of writing. (Note: it is not possible to give these figures for SQL-based relational products because the language works in a non-procedural way—the number of instructions and disk accesses required to execute a SELECT will depend on the query.)

Typical timing figures for CODASYL and hierarchical databases are:

No. of instructions/DBMS call	~2000
No. of application program instructions/DBMS call	~1000
No. of instructions/message pair	~1000
Disk access time	~30 ms

For a 1 MIP machine (1 million instructions per second) this gives:

Time/DBMS call	2 ms
Time application program/DBMS call	1 ms
Time/message pair	1 ms

These timing forms are filled in using the Physical Process Specification as a guide, particularly the No. Acc column though in this case we have revised the figures by specifying the 'from' and 'to' sets linking Booking to Depot as sorted sets. The page planning charts developed for the space sizing are also needed as these tell us which RECORDs are on the same page and can therefore be retrieved with a single disk access. The columns under the Disc Transfers on the timing form shown estimates of the number of disc reads and writes for each RECORD type retrieved. Note that for a write two transfers will be required; one to read in the page and another to write it back. If an index-based DBMS was used (many relational products use indexes) then columns should be added to the form to show the disc transfers associated with index reading and writing. Throughout the tuning phase these timing forms may need to modified many times as the performance is recalculated after each change in design. One way of making this process of recalculation less painful is to set up these timing forms on spreadsheet packages so that recalculation can be performed automatically.

Codasyl transaction timing form

Transaction:	Daily Departures and Returns List	No.	1	Frequency:	daily	Batch / On-line

Process Verbs:	F - FIND	C - CONNECT	Read Path:	C - CALC
	G - GET	D - DISCONNECT		N - NEXT pointer
	O - OBTAIN	E - ERASE		P - PRIOR pointer
	M - MODIFY	S - STORE		O - OWNER pointer

RECORD	DBMS		No. Dbms calls per trans	Disk Transfers			CPU time				Comments
	Process Verb	Read Path		Read	Write	Tot	DBMS	TP Mon	App Prog	Tot	
Local Office / Depot	O	N	51	51		51	102		51	153	Read 50 in set + 1 to ensure all found
Booking	O	N	50 x 70 = 3500	3500		3500	7000		3500	10500	Set is sorted in departure date order only 1050/7 need to be read
Customer	O	O	50 x 8 = 400	400		400	800		400	1200	
Driver	O	O	50 x 6 =300	300		300	600		300	900	
Booking	O	N	50 x 70 = 3500	3500		3500	7000		3500	10500	Set is sorted in return date order only 1050/7 need to be read
Customer	O	O	50 x 8 = 400	400		400	800		400	1200	
Driver	O	O	50 x 6 =300	300		300	600		300	900	

	Total No. disk transfers:	8451	Total CPU time:	25353	
	Total I/O time:	253350 s	Total resource time:	278.884 s	

Fig. 8.14

Accurate timings for transactions run on relational DBMSs are often more difficult to estimate than for other DBMS products or conventional file handlers. This is because the physical access navigation and retrieval mechanisms are controlled by relational DBMS software. Thus without a very good knowledge of the internal workings of a particular product it is almost impossible to predict the performance. It is more accurate and much easier to build and populate a prototype database and then run the SQL for the critical transactions on it. The timings for the full system can then be estimated by extrapolation from the prototype timings.

Estimation of recovery times
It is difficult to predict how often system failures will occur. Suppliers may quote and contractually guarantee failure rates for hardware but it is harder to predict software and human errors. What is needed is planning for the various types of failure in the knowledge that they will occur.

The times taken for recovery should be estimated for the following types of failure: system crashes (e.g. power failure), disk failures (e.g. head crash), and program or transaction failures (e.g. overflow error). These estimates will be based upon the following factors:
- frequency of dumping;
- DBMS facilities for restoring, checkpointing, and transaction logging;

- extent to which these facilities are used;
- acceptable errors or missing data (determined by the users).

In case of prolonged loss of the system (e.g. through fire), the performance of back-up manual and/or stand-by computer sites should also be estimated and measured against the users' objectives.

Improving performance

Initially the performance of the First Cut Physical Data and process designs are estimated and compared with the Performance Objectives. If these objectives are not met then these designs will need to be examined and improvements made. After each modification the performance of the system, in all areas likely to be affected by the change, should be recalculated. When the performance is meets the objectives the project can proceed to full development.

It is important to realize that making improvements in one area of performance may lead to a degradation in another area. For instance, a reduction in the storage space made by removing some indexes will slow down those transactions which used the indexes.

Below we give some brief hints on how performance can be improved in storage space, timings, and recovery. System software will vary greatly in the extent to which these are possible. Generally a detailed knowledge of the system software and hardware is required to get optimum performance from the technical environment. The project may choose to buy expertise in the form of consultancy from, usually, the software supplier.

Reducing the storage space requirements

This means operating on the First Cut Physical Data Design and the Logical Design Volumes. Some possible reductions can come from:

- codifying data;
- compressing data, e.g. dates as 2 bytes, A/N as 5 bits, removing trailing spaces;
- archiving earlier;
- dropping rarely used data;
- relying on data in other systems;
- reducing the number of pointers and indexes;
- removing redundant or derived data held for timing reasons;
- removing duplicated foreign keys in non-relational databases

Speeding up transactions

There are two major ways to improve the speed of transactions. First, by modifying the program design to make it more efficient, and secondly by modifying the the data design to make retrieval faster. It is always preferable to take the first approach if at all possible. Modifying the data structure will have effects on the performance in all areas, including other timings.

Some possible ways to speed up transactions are suggested below:

- ensure access paths in programs are efficient;
- modify clusters by adjusting allocation to pages to reduce disk accesses;
- introduce more access methods, e.g. extra pointers, indexes;
- drop indexes before batch updates and recreate afterwards;
- modify the depth and structure of B* indexes;

- unnormalize data in relational databases;
- use read-only snapshots to reduce locking overheads;
- perform functions in a different (faster) way than that specified by the user;
- introduce new data groups (not on the data design) to reduce access paths;
- increase buffer space;
- reduce the frequency of dumping and checkpointing.

Improving recovery

Recovery times can be improved by increasing the frequency of dumps, by increasing the frequency of checkpointing, and by maintaining transaction logs, before-image and after-image logs. Note that these will increase both the space requirement and the transaction timings.

Changing objectives

If it becomes apparent that the Performance Objectives set in stage 3 cannot be met, then a case for changing them is presented to the users who originally set the objectives. The development team should prepare a list of possible changes and present them to the users. The sort of changes that might be suggested are:

- reductions in the amount of historical data held;
- increased response times or batch run times;
- increased hardware performance requirements (i.e. spend more money);
- a reduction in the functions offered by the system

Changes to the Performance Objectives should really be regarded as a last resort rather than a normal course of action.

SUMMARY

- Performance Objectives are set in stage 3 for space, timing, information, and recovery.
- The performance of the First Cut Physical Data and process design is estimated and compared with the objectives.
- Improve performance to meet objectives by modifying the process and data designs.
- After each change the performance is estimated.
- The final design will usually be a compromise between objectives.
- If objectives cannot be met then changes to them are agreed with users.
- When the performance is acceptable the project can proceed to full development.

8.5 Preparing for the implementation (steps 640–690)

After the system has been tuned and an acceptable design has been reached the documentation required for construction and the plans for the remaining phases can be finalized. Much of the activity in these final steps involves drawing together documentation produced earlier in the project. There are no new SSADM techniques employed in these steps and most of the work falls into the category of traditional system design. SSADM gives some guidelines and suggests the end-products for these activities, but many projects find these are covered by their installation's or organization's standards. As in the rest of stage 6, the approach taken will also depend greatly on the hardware and software environment.

The diagram of stage 6 (Fig. 8.1) suggests that all these steps are carried out concurrently. There are some dependencies between the steps but they are largely independent of each other. Many of the activities involved will have been considered, and some started, earlier in the project and particularly in the technical options stage.

In this section we discuss each of the steps briefly. More general and more detailed information on some of the topics addressed can be found in the books listed in the bibliography.

Step 640: create file/database definitions

Once the design has been tuned to acceptable performance, the file or database definitions can be loaded into the data dictionary or system library. This step involves taking the First Cut Physical Data Design as modified by step 630 and producing data definition statements. This was discussed in Sec. 8.2 and an example was given of SQL PLUS statements to define the Yorkies database.

Step 650: complete program specifications

Only partial Program Specifications for critical programs were developed before tuning. Full Program Specifications are developed for all programs in the new system. The format of the Program Specification may be defined by installation standards. If fourth-generation software is to be used, the specifications will need to be geared to the requirements of the particular package.

All critical transactions were originally specified before tuning, and probably modified during tuning. In this step further detail is added to enable them to be used as Program Specifications for the construction phase. All of the remaining non-critical programs are also specified completely. The techniques used for this specification are those used in step 620, *create program specifications for major transactions*, which were discussed in Sec. 8.3. Below we give a checklist of items required for each program.

Documentation required for all programs:

- Physical Program Description.
- Physical Process Specifications.
- Specification of all input and output formats.
- Specification of error and system messages.
- Error conditions under which transactions may be abandoned.
- Specifications of files/records/data items accessed.
- List of common facilities/routines to be used.
- Timing forms for each critical transaction.

For on-line programs:

- Dialogue specification (usually Logical Dialogue Outlines and Controls).
- Help facilities.
- Security restrictions.

For batch programs:

- System run chart showing all processing cycles.
- Batch run chart showing where the program fits into its processing cycle.

For each program a detailed test plan is created with test transactions and expected results identified. Responsibility for testing is allocated and a schedule of test runs is created.

Also as part of this step detailed plans for the construction phase are created. These will include the sequence of program construction, incorporation of programs into sub-systems, and plans for program testing. Resource plans and elapsed time schedules are created. It is suggested that the products of this step are reviewed formally.

Step 660: create systems test plan

The testing strategy and rough test plans can be developed throughout the design phases of the project. However, it is only at this time that detailed plans can be created.

The suggested testing strategy is to follow the conventional sequence of integration, system requirement, acceptance, and volume testing. For each of these it is necessary to allocate responsibilities for testing, to define test cases, and to specify expected results. A detailed elapsed time schedule is developed for the testing phase; this cannot be finalized until programming is complete.

Step 670: create operating instructions

Operating schedules are created for each batch processing cycle in the system. These, and the operators' instructions for on-line programs, will form the basis of the operations manual.

Step 680: create implementation plan

Plans for the changeover will have been considered in the selection of technical options (stage 3) and possibly earlier in the project. However, once the test plans have been completed, detailed data conversion, changeover, installation, and other resource plans can be developed.

The approach to this step is usually defined by project management standards or methods. It should be emphasized that these issues of conversion, changeover, and installation are critical to the success of the overall project. In some cases each of these should be approached as a large project in its own right.

Step 690: define manual procedures

The manual procedures that interface with the automated system need to be designed. Some projects extend the Data Flow Diagrams outside the boundary of the automated system. They can then be supported with further narrative descriptions to define the manual procedures. This approach has the advantage that users and analysts will both be familiar with the technique although traditional clerical flowcharts are often a better way of specifying manual systems.

The specification of manual procedures will go into the user manual together with the following information:

- Formats of all screens, reports, and input forms.
- Definition of all on-line dialogues.
- Definition of input methods and output distribution for batch programs.
- Operating schedules.

- Instructions on the use of back-up and recovery facilities.
- Instructions on the use of end-user facilities such as query languages.

This step will also involve the development of plans for training the users.

Appendix A. SSADM Glossary

Access path The route through a data structure that must be navigated to satisfy an enquiry or update.

Application generator A software tool that allows applications to be implemented more rapidly than conventional programming.

Attribute See data item.

Batch The mode of processing that requires no intervention from a user from beginning to end, using a 'batch' of information.

Batch function A function of the system that will be implemented as a batch program in the required system.

Bottom-level Data Flow Diagram A Data Flow Diagram that is not further decomposed to a lower level.

BSO See Business System Options.

Business System Options A set of options, normally in the form of top-level Data Flow Diagrams, that allow the user to decide upon how the system will impact on the business of the organization.

CLDD See Composite Logical Data Design.

Context Diagram A single process representing the entire system under investigation. Drawn for the purpose of examining the interfaces between the system and external source/recipients.

Common Processing Catalogue A means of cataloguing any duplication of processing within the system e.g. operations appearing on more than one process outline, library programs that exist in the common system or entries in the Elementary Function Descriptions that are the same for different bottom-level Data Flow Diagram processes.

Composite key The unique identifier of a relation, composed of more than one element. At least one of the elements would not have a unique value in isolation from one or more of the other elements.

Composite Logical Data Design A data structure derived by the combination of the Required System Logical Data Structure and the data structure resulting from relational data analysis.

Compound key The unique identifier of a relation composed of more than one data item. Each data item has a unique set of values in isolation from the other elements.

Data Catalogue A catalogue of all data items referenced in any of the SSADM documentation, giving all relevant size, validation, and descriptive information about the data items.

Data flow An element on a Data Flow Diagram representing the flow of information to or from a Data Flow Diagram process. A data flow at one level of a Data Flow Diagram may be decomposed to several data flows at lower levels.

Data Flow Diagram A diagram representing the flow of information around a system, the way in which it is changed and stored, and the sources and recipients of the information outside the system.

Data group A group of data items uniquely identified by a key. Data groups appear on the Composite Logical Data Design and are based upon the entities from the Logical Data Structure and the relations from the relational data analysis.

Data Inventory See Data Catalogue.

Data item The smallest unit of information in a system. Data items are grouped into entities or data groups on data structures and are shown to be flowing around a system on data flows.

Data store A repository of information within a system. For main data stores, each store represents a whole number of entities or data groups from the Logical Data Structure or Composite Logical Data Design. Transient data stores are temporary stores which accumulate information before it is used and subsequently deleted.

Data structure See Logical Data Structure, Composite Logical Data Design, and relational data structure.

DFD See Data Flow Diagram.

Dialogue A set of exchanges between the user of a system and the system in an interactive way, resulting in a retrieval or update of information.

Effect The particular set of changes caused within an entity by an event. It is represented on an Entity Life History as a box with no subsidiary boxes. A state indicator value is set by an effect box, and a set of valid previous state indicator values can be defined as part of the validation criteria for the effect.

Elementary Function Description A brief, normally textual, description of a bottom-level process on a Data Flow Diagram. It is used to understand the Data Flow Diagrams and not as a basis for design.

ELH See Entity Life History.

ELH Matrix The start-up document for Entity Life Histories. The entities from the Logical Data Structure are listed on one axis of the matrix and events (initially identified from the Data Flow Diagrams) listed on the other axis of the matrix.

Entity A thing about which the system will be required to hold information. An entity appears on the Logical Data Structure. When the Composite Logical Data Design is constructed, the entities form the basis for the data groups.

Entity/Data Store Cross Reference A standard SSADM form used to indicate the relationships between the entities on the Logical Data Structure and the data stores on the Data Flow Diagrams. This is a one-to-many relationship for the logical and required system definitions.

Entity Description A standard SSADM form used to describe each entity (or data group) in more detail. It consists of a textual description and a list of data items belonging to the entity.

Entity Life History An Entity Life History is a diagrammatic representation of the life of a single entity, from its creation to its deletion. The life is expressed as the permitted sequence of events that can cause the entity to change. The components of the notation are sequence, selection, iteration, quit and resume, and parallel structures. State indicators are added to the diagram.

Entity role If a single event occurrence affects different occurrences of an entity, but at different points within the Entity Life History, the entity is being affected in different roles. (For example, one entity occurrence may replace another occurrence—one is created at the same time as the other is deleted).

Entry point The point in the Logical Data Structure or Composite Logical Data Structure where the accessing required by an event or enquiry begins.

Events An event is whatever brings a process into action to change entities. Although it is a process that performs the change, it is the event that is the cause of the change. The permitted sequence of events within the life of an entity is modelled using an Entity Life History.

Event Catalogue A standard SSADM form used to list all events, describe them, and indicate which entities are affected by each one. Volumes may be added to the form to help in an initial sizing of the system.

Exchange Part of a dialogue between the user and the computer system. A single exchange consists of a single input from the user and the corresponding reply from the system.

Exclusive arc A device on a data structure diagram to indicate the mutual exclusion of a set of relationships.

External entity See external source/recipient.

External source/recipient Whatever or whoever donates information to or receives information from the system. Represented on a Data Flow Diagram as an oval.

Feasibility study A brief study into the system before the start of a full SSADM project to determine its feasibility. This can be done using the feasibility study guidelines based upon the first three stages of SSADM.

Feasibility Study Report The output from a feasibility study. This may be relatively detailed in terms of the system requirements and is used as a basis for planning the project.

First Cut Physical Data Design The physical data design is an executable design specific to the file handler or database management system. The First Cut Physical Data Design is the first version produced by applying simple rules, specific to the file handler or database management system, to the Composite Logical Data Design. These simple rules are known as the first cut rules. The First Cut Physical Data Design is usually represented diagrammatically.

First cut rules Simple rules applied to the Composite Logical Data Design that turn it into the First Cut Physical Data Design. The first cut rules will be specific to a particular software product. However, the rules are very similar within classes of software products. Thus all relational database management systems will have similar first cut rules.

Foreign key Data item(s) appearing in an entity, relation, or data group that are also the primary key of another entity, relation, or data group. Foreign keys are identified during relational data analysis. They indicate the 'many' end of a one-to-many relationship on data structures.

Function Update functions are identified from the Required System Data Flow Diagrams and are denoted as being either batch or on-line. A top-level process represents one or more functions. If a top-level Data Flow Diagram process consists of entirely batch processes that occur with the same regularity (e.g. weekly, monthly, etc.) it is documented as a single batch function. Similarly, each top-level process that is entirely on-line becomes a function. Otherwise, the second-level Data Flow Diagrams are inspected. Batch processes are separated from on-line processes, and those that occur with the same frequency are grouped to form functions. The on-line processes are grouped to become on-line functions. The functions are documented on separate on-line and batch Function Catalogues and are later used as a basis for program design.

Only very major enquiry functions will be represented on the Data Flow Diagrams. Otherwise, the required enquiries should be documented directly as functions on the Retrievals Catalogue. It is not necessary initially to distinguish between batch and on-line enquiry functions, as their implementation may be entirely dependent upon the size of the report produced rather than some explicit requirement for batch or on-line implementation.

Function Catalogue A standard SSADM form used to describe functions.

I/O Description A standard SSADM form used to catalogue the data items that appear on inputs and outputs of the system. Can represent screens, reports, etc.

LDC See Logical Dialogue Control.

LDO See Logical Dialogue Outline.

LDS See Logical Data Structure.

Leg (parallel structure) Part of an Entity Life History. When more than one event, or group of events, can affect an entity in parallel, they are placed under a double bar to denote this parallellism. Each group that is placed under the parallel bar is a separate leg of the parallel structure.

Logical Data Flow Diagram A Data Flow Diagram reflecting the functions of the current system without the physical constraints. The Current Physical Data Flow Diagram is converted into the Logical Data Flow Diagram as the first step of specifying the required system.

Logical data group See data group.

Logical Data Structure A diagram showing the underlying structure of the data either for the current or required system. The objects on a Logical Data Structure are principally entities and relationships. All relationships are one-to-many. This is a very central technique of SSADM. The Required Logical Data Structure is used as the basis for the Composite Logical Data Design.

Logical Design Volumes The number of occurrences of data groups and relationships on the Composite Logical Design. These volumes are added towards the end of stage 4 to be used as a basis for the file or database design and sizing. The volumes are documented on a standard SSADM form.

Logical Dialogue Control A high-level Logical Dialogue Outline showing the relationship between two or more event-level dialogues.

Logical Dialogue Outline A standard SSADM form used to show the exchanges for an on-line event or enquiry. The form is used as a form of paper-based prototyping to agree the dialogues with the user.

Logical Enquiry Process Outline A standard SSADM form detailing the operations and accessing required to satisfy a single enquiry.

Logicalization The procedures to convert a Current Physical Data Flow Diagram into a logical view of the same system (the Logical Data Flow Diagram).

Logical Update Process Outline A standard SSADM form detailing the operations and accessing required to satisfy a single event (or sub-event). Based upon the information in the ELH Matrix and Entity Life Histories. Each entity affected by an event is listed on the Process Outline, together with any entities that require to be accessed.

LEPO See Logical Enquiry Process Outline.

LUPO See Logical Update Process Outline.

Navigation path See access path.

Normalization See relational data analysis.

Operation The specification of a unit of processing that is performed as part of a Process Outline. This is the most detailed definition of processing reached in logical design. A single operation may appear on more than one Process Outline.

Operational master A data item in an entity, which is not part of the key of that entity, on which direct access is required. This is shown as an oval owning the entity on the Logical Data Structure. The data item name is written inside the oval

Optimization The last step of relational data analysis. This involves the comparison of all relations that possess the same key in order to merge them. The result represents a fully normalized set of relations.

Performance Objectives These are developed in step 340 after the technical options have been decided. The users, assisted by the development team, define objectives for data storage, timings for critical transactions, recovery, and information.

Physical design Stage 6 of SSADM. This involves the conversion of the detailed logical design produced as a result of stages 4 and 5 and converting it into a fully workable, implementation-dependent design.

Physical design control The procedures to convert a first-cut design into a tuned design that would meet the Performance Objectives set by the users.

Physical Document Flow Diagram A start-up to the production of the Current Physical Data Flow Diagrams. This approach is appropriate if the current system consists principally of flows of information in the form of documents or computer input and output. Each source or recipient is represented as an oval in a diagram with the documents represented as flows between them

Physical Goods Flow Diagram See Physical Resources Flow Diagram.

Physical Resources Flow Diagram A start-up to the production of the Current Physical Data Flow Diagrams. If a system consists primarily of flows of goods rather than flows of information, the current system may be modelled by the flows of goods rather than the flows of information.

Physical Process Specification The implementation-dependent equivalent of a Process Outline. This is a standard SSADM form.

Physical Program Description A front page added to the Program Specification to summarize the contents and list the relevant documents. A standard SSADM form.

Primary key A data item or combination of data items (see compound key) whose value(s) uniquely identify the values of other data items in an entity, relation, or data group. Primary keys are identified for all relations during relational data analysis. During development of the Logical Data Structure it is often useful to identify notional primary keys before they are formally identified by relational data analysis.

Problems/Requirements List One of the most central documents in the analysis stages of SSADM. It is used as a check-list of all the factors that must be accounted for in the new system and can be used to measure the success of a project by checking that all of the problems and requirements have a corresponding solution. A standard SSADM form.

Process An object on a Data Flow Diagram showing the transformation of input data flows to output data flows.

Process Outline See Logical Enquiry Process Outline and Logical Update Process Outline.

Program Specification A collection of documents, including the Physical Program Description and several Physical Process Specifications, which is given to the system developer to implement.

Project A distinct development leading from the initial requirement for a new system to its implementation.

Project control The process of ensuring the project is running as planned, within timescales, budgets, and objectives. Project control methods specify procedures for monitoring and reporting project progress. Also, various automated tools provide support for project planning.

QA See quality assurance.

Quality assurance The process of ensuring the products of a project are of a sufficiently high standard.

Quality assurance review A formal meeting of users and technical reviewers held normally at the end of each stage of SSADM. In the review, the products of the stage are reviewed to ensure that they are consistent with the SSADM technical standard and meet the requirements of the users.

Quit and resume A device on an Entity Life History to indicate a departure from the sequence specified in the diagram.

RDA See relational data analysis.

Relation A group of non-repeating data items identified by a unique key. Derived as a result of relational data analysis.

Relational data analysis The process of normalizing data items to exclude duplication or repeating items. Normally performed on groups of data items from the required system such as I/O Descriptions or screen definitions. A series of steps is followed going from an unnormalized form through first, second, and finally third normal form. The relations derived as a result of the third normal form are combined as a step of optimization.

Relational data structure A data structure based upon the relations resulting from relational data analysis.

Relationship A logical connection between entities on a data structure. Generally, only one-to-many relationships are permitted. The relationship is denoted by a line with a 'crow's foot' to indicate the 'many' end of the relationship.

Required System Data Flow Diagram The Data Flow Diagram developed to support the requirements of the new system.

Retrievals Catalogue A Function Catalogue referencing enquiry-only functions.

Software support tool A computer-based tool designed to support a method, either in terms of diagramming assistance or as a central repository of development information. Often referred to as a CASE (computer-aided systems engineering) tool.

SSADM Structured Systems Analysis and Design Method. The standard UK government method for the analysis and design of computer systems.

SSADM documentation A series of forms, each with a specified purpose within the method.

Stage SSADM consists of six stages, each of which is further defined in terms of steps, inputs, and outputs. An SSADM project will move sequentially through the stages with the exception of stages 4 and 5 which are done in parallel.

State indicator A state indicator may be thought of as a data item within an entity that is updated each time an event affects the entity. The updating of a state indicator is dictated by the Entity Life History structure. Each effect on an Entity Life History is bounded by a set of values of the state indicator that are valid prior to the effect and a single value that is set as a result of the completion of the effect. The 'valid previous' values of the state indicator are reflected in the Logical Update Process Outlines as validation criteria on the operations and the 'set to' value is included in the updating operations. Subsidiary state indicators may be introduced to control the different legs of a parallel structure within Entity Life History diagrams.

Step A sub-division of a stage of SSADM. Each step is defined in terms of inputs, outputs, and tasks.

Sub-event If two distinctly different paths of processing might result from the receipt of a single triggering event, and the choice of processing path is determined only after the receipt of the trigger, the two different reactions of the system are classified as sub-events and catalogued separately on the Event Catalogue, except to indicate what is required to differentiate between them.

Subsidiary state indicator A subsidiary state indicator may be thought of as an additional data item in the entity that is updated by one leg of a parallel structure. If it is necessary to track each leg of a parallel structure on an Entity Life History, a separate subsidiary state indicator is used for each leg.

System Whatever is within the boundary of a study.

Technical options Stage 3 of SSADM allows the assessment of possible technical solutions based upon the detailed analysis performed in stage 2. These options are based upon the requirements and constraints stated in the analysis. Each option is described in detail to the users, including a cost/benefit analysis, and the user is then helped to make a decision as to which option to choose for implementation.

Terms of reference The statement of the work the developers are asked to undertake in an SSADM project. This should be present at the very beginning of a project.

TNF See relational data analysis.

TNF Data Structure See relational data structure.

Transaction A physical equivalent to an event or enquiry.

User A term used to encompass the end-user of the required system, managers in the user organization, or customers for the system. The user provides the requirements of the system, is responsible for funding the system, and signs off each stage before progressing to the next.

User options See technical options and Business System Options.

User Requirement A statement of the requirements of the new system, normally provided entirely by the user. This will form the basis for the Problems/Requirements List.

Volumetrics The frequencies and volumes of various elements of the specification and design. For example, average and maximum volumes of events may be added to the Event Catalogue. Also volumes and relative volumes of data groups shown on the Composite Logical Data Design are documented by Logical Design Volumes.

Appendix B.   Bibliography

Avison, D.E. and Fitzgerald, G. (1988) *Information Systems Development: Methodologies, Techniques and Tools* (Blackwell Scientific, Oxford).

Curtice, R.M. (1987) *Strategic Value Analysis: A Modern Approach to Systems and Data Planning* (Prentice-Hall, Englewood Cliffs, NJ).

Daniels A. and Yeates, D. (1988). *Basic Systems Analysis,* 3rd Edn (Pitman, London).

Date, C.J. (1986) *An Introduction to Database Systems: Volume 1,* 4th Edn (Addison-Wesley, Reading, Mass.).

DeMarco, T. (1979) *Structured Analysis and System Specification* (Prentice-Hall, Englewood Cliffs, NJ).

DeMarco, T. (1982) *Controlling Software Projects: Management, Measurement, and Estimation* (Prentice-Hall, Englewood Cliffs, NJ).

Galliers, R. (Ed.) (1987) *Information Analysis: Selected Readings* (Addison-Wesley, Reading, Mass.).

Jackson, M.A. (1975) *Principles of Program Design* (Academic Press, London).

Keen, J. (1987) *Managing Systems Development,* 2nd Edn (John Wiley, New York).

Longworth, G. and Nicholls, D. (1987) *The SSADM Manual* (National Computer Centre, Manchester).

McFadden, F.R. and Hoffer, J.A. (1988) *Data Base Management,* 2nd Edn (Benjamin/Cummings, Menlo Park, Calif.).

Olle, T.W., Hagelstein, J., Macdonald, I.G., Rolland, R., Sol, H.G., Van Assche, F.J.M. and Verrijn-Stuart, A.A. (1988) *Information Systems Methodologies: A Framework for Understanding* (Addison-Wesley, Reading, Mass.).

Shneiderman, B. (1987) *Designing the User Interface: Strategies for Effective Human Computer Interaction* (Addison-Wesley, Reading, Mass.).

Sommerville, I. (1989) *Software Engineering,* 3rd Edn (Addison-Wesley, Reading, Mass.).

Yeates, D. (1986) *Systems Project Management* (Pitman, London).

Appendix C.   Yorkies case study

Terms of reference

1.   To design a computer system to support the vehicle rental, driver administration, customer records, and invoicing areas of Yorkies.
2.   To investigate ways of improving the efficiency of the operations of the company in those areas specified in 1.
3.   To investigate extending the system to include the administration of one-way hires and the acceptance of non-registered customers

Background information

Yorkies are a medium-sized organization dealing with the hire of vehicles ranging in size from vans to articulated lorries. They have a Head Office (HQ) and 50 Local Offices with a depot attached to each Local Office. They deal with a large number of regular local customers over the whole of the UK mainland. The organization chart for Yorkies is shown in Fig. C.1.

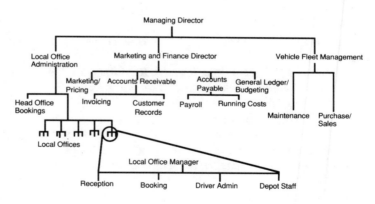

Fig. C.1

Description of the current system

1.   Customers may make telephone or written bookings of vehicles to reception staff in the Local Offices. These are passed to the booking staff where they are first checked against the local Customer List to ensure the customer is accredited and that the booking does not exceed the maximum value bookable for that customer. If the booking is unacceptable the customer is referred to HQ.

2.   If the booking is satisfactory the Vehicle Booking Diary is checked to ensure that a suitable vehicle is available for the days required; if no vehicle is available the

unfillable booking is passed to HQ who use the Empty Vehicle Log to find an office that can satisfy the booking.

3. If a vehicle is available a Booking Sheet is partially filled out and the booking number and the vehicle category entered into the Vehicle Booking Diary. The Booking Sheet is in four parts; the bottom, partially filled copy, is sent to the customer as confirmation of his order. Any instructions to the driver are filed in order of the date required and stored separately until required. If a driver is required the driver administration staff are notified.

4. The Empty Vehicle Log is amended by HQ every second working day from sheets filled out by the Local Offices detailing the number of vehicles of each category currently available for hire for each of the next 20 working days.

5. Any previously unfilled bookings filled by HQ using the Empty Vehicle Log are passed to the Local Office able to satisfy the booking. They fill out a Booking Sheet and deal with the booking in the normal way. If HQ fill the booking using the Empty Vehicle Log they inform the customer directly.

6. If a driver is required the Local Office find a suitable driver from the Driver/Agency Register held at each Local Office and contact him by telephone to determine whether he is available for the days required. When a driver is 'booked' the booking details are entered into the Driver Register and the driver name onto the Booking Sheet. The driver instructions on the Booking Sheet are sent to the driver as confirmation. If no driver can be found the Local Office contacts an agency who supply drivers.

7. When the vehicle is collected by the customer or driver the date and time collected, and departure mileage are added to the Booking Sheet and to the Vehicle History Card. The booking number is also added to the Vehicle History Card. The activities concerned with vehicle departure and return are performed by the depot staff.

8. When the vehicle is returned the mileage, time and date, and vehicle condition are entered onto the Booking Sheet. The completed Booking Sheet (copy 2) is passed to the customer and a copy filed at the Local Office. The top copy is sent to HQ where it is filed and used for invoicing. The return details are also entered onto the Vehicle History Card. If the vehicle is damaged then this is recorded with the vehicle condition. If a driver was supplied the driver hours are entered in the Driver Register. If the vehicle goes out of or comes into service this is notified to the booking staff and written on the History Card and notified to fleet maintenance at HQ.

9. The accounts section is based at HQ and uses a microcomputer accounts package to produce invoices and to hold customer records. This uses two files: the Customer File and the Invoice File.

10. The Customer List is printed out monthly and sent to each Local Office. Every week a list of amendments are printed and sent to each Local Office. Maintenance of the Customer File (e.g. adding new customers, changing their details, etc.) is carried out by HQ accounts staff.

11. Copy A of the completed Booking Sheet is sent by the depot staff to HQ accounts and is used to produce the invoice. Customers are invoiced on individual bookings but also

receive monthly statements and reminders if they have not paid in the last 30 days. Completed Booking Sheets are filed for archival.

12. When payment is received the remittance is recorded and reconciled against the invoice in the Invoice File.

13. The amount owing at the end of each month is recorded on the Customer File—this information is also on the monthly statement sent to the customer.

Information currently held

Customer List (Local Offices) The invoicing of customers is carried out centrally at HQ using a microcomputer which produces monthly a list of accredited customers that is sent to each of the Local Offices who use it when accepting and confirming bookings. It contains customer number, customer name and address, and a maximum value bookable. About 500 customers are dealt with on a regular basis. Weekly amendments to this list are sent out by HQ.

Driver Register (Local Offices) This is used when Yorkies are requested to supply a driver with the vehicle. It contains the driver's name, address, telephone number, and the type of licence held (HGV I, II, or III). Bookings for the driver are entered with the booking number and dates required. When the vehicle is returned the hours worked are added.

Booking Sheets (Local Offices and HQ) The Booking Sheet, when completed, contains the following information: office issuing, booking number, booking date, customer number, customer name and address, category of vehicle required, vehicle made and model, vehicle registration number, date and time required, date and time collected, mileage when collected, date and time to be returned, date and time when actually returned, mileage when returned, vehicle condition, driver name, driver instructions.

Vehicle History Cards (Local Offices) Each office keeps a set of cards for each vehicle it is responsible for. They are filed by make and model within category (e.g van, articulated lorry, etc.). The following information is held: make and model, vehicle category, registration number, date purchased; and information relating to each hiring the booking number, date and mileage when sent out, date and mileage when returned, the driver name (if Yorkies supplied the driver), and any comments about the condition. Information about the various vehicle categories is kept with the History Cards.

Vehicle Booking Diary (Local Office) This is used when acceptance or rejecting bookings. Each day that a vehicle (and driver, if required) is booked is recorded.

Empty Vehicles Log Diary of unbooked vehicles kept at HQ and used to fill bookings that could not be filled by the Local Office that they were originally submitted to. This is kept 20 working days forward and for each day holds for each office the number of vehicles of each category that are currently unfilled.

Customer File Microcomputer file held at HQ which holds customer number, name and address, telephone number, maximum amount bookable, and amount owing at end of previous month.

Invoice File Microcomputer file held at HQ which holds invoice number, customer number, customer name and address, booking number, cost of booking, date payment received, amount paid, date of booking, and date of invoice.

Current problems

- Customer List held at Local Offices is frequently out of date. This causes customers to be wrongly accepted or rejected.
- Current microcomputer invoicing system is very heavily overloaded and could not handle any expansion in the business. It is a single-user system that uses an invoicing package and it is not possible to upgrade it to a multi-user system.
- It is not straightforward to accept 'off-the-street' customers.
- Customers cannot be informed immediately whether vehicles are available for them. If a booking has been referred to Head Office it may take several telephone calls to check on its progress.
- The summary Booking Sheets created by Head Office take a long time to prepare and are often inaccurate. This causes booking requests to be sent by HQ to the Local Offices for vehicles already booked or out of service.
- Expensive agency drivers are often used in some offices when other offices may not make full use of the drivers on their register.
- There is no standard way of dealing with cancelled bookings, each Local Office has developed its own procedure.
- It is difficult, with the current accounting procedures, to calculate the relative profitabilities of the Local Offices.
- Statistical reports are often late and require considerable clerical effort for their production.

New requirements

1. Ability to deal with one-way hires and to track vehicles.
2. Have a central pool of drivers organized regionally who are allocated to vehicles.
3. Investigate the possibility of setting up regional offices to administer groups of Local Offices.
4. Extend the business to allow one-off customers to hire vehicles for removals, etc.

Volumetrics

- Currently about 3000 customers are dealt with per year, although there are 5250 on the Customer List. This is growing at the rate of 5 per cent per year.
- There are currently 50 Local Offices operating, this has increased recently due to a company merger, but is expected to remain constant in the foreseeable future.
- An average of 50 vehicles per Local Office—2500 in all.
- There are currently 61 vehicle category codes in use. This is not expected to change.
- An average of five permanent drivers are attached to each office. Each Local Office register lists about 50 drivers.
- About 7000 bookings are accepted every year by each Local Office

Appendix D. Suggested answers

3.3

Figure D.1 shows the Current Physical Data Flow Diagram for the CAt Breeding Agency and Fig. D.2 shows the Current Physical Data Flow Diagram for Reckitt Repairs.

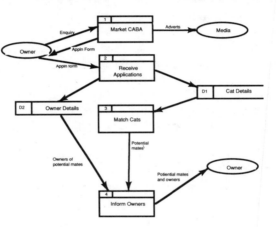

Fig. D.1

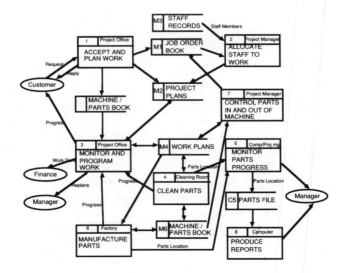

Fig. D.2

3.4

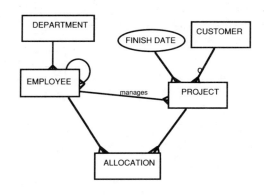

Fig. D.3

Figure D.3 shows the Logical Data Structure for the Scapegoat Systems project management system. The recursive relationship is shown on Employee because each employee is allocated one manager within the department. Note that an additional 'one-to-many' relationship is required between Employee and Project to show that one employee can manage many projects but that each project can have only one project manager.

The relationship between Customer and Project is optional because some projects are internal and are therefore are not done for a customer.

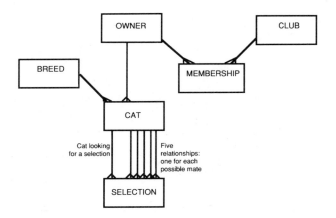

Fig. D.4

Figure D.4 shows the Current System Logical Data Structure for CABA. This is a tricky exercise. The difficulty is in deciding how to represent the selection itself. Our solution uses a separate entity which in effect represents the information about the selection contained on the cat card. This Selection entity has one master occurrence for the cat on the application and five master occurrences for the five possible mates. Another way to represent this would be to represent the relationship between Selection and Cat as a 'many-to-many' and create a link group. However, this would be less accurate as precisely five cats are recommended, but it would be more flexible in that if CABA wanted to offer six or more mates it would be easier to make the change.

Figure D.5 shows the Reckitt Repairs Logical Data Structure

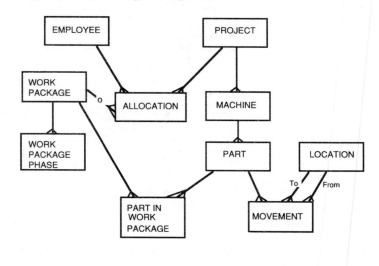

Fig. D.5

4.2

1. Logical Data Stores

Data store		Entity
D1	Projects	Project
		Allocation
D2	Machine Details	Machine
		Part
D3	Location	Location
D4	Movements	Movement
D5	Employees	Employee
D6	Work Details	Work Package
		Part in Work Package
		Work Package Phase

2. Redundant processes

In processes 4–7 there is duplication of processing in that the movements are monitored twice—by the current computer system and by the manual system in the factory and cleaning room.

The actual cleaning and manufacturing are outside the system boundary as these are not easily automated and the robotics required are outside the scope of the present project.

3. The Reckitt Logical Data Flow Diagram is shown in Fig. D.6

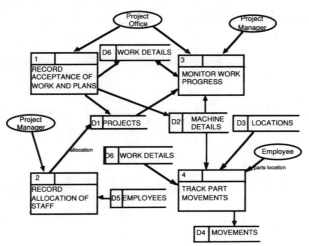

Fig. D.6

4.6

Figure D.7 shows the Scapegoat Systems Required System Logical Data Structure.

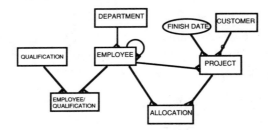

Fig. D.7

4.7

Figure D.8 shows the Entity Life History for the Project entity of Scapegoat Systems.

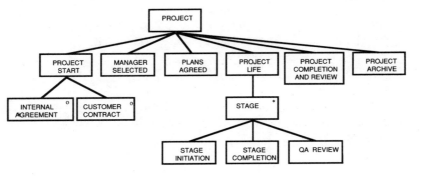

Fig D.8

4.8

The Logical Dialogue Outline for the event New Employee in Scapegoat Systems is shown in Fig. D.9

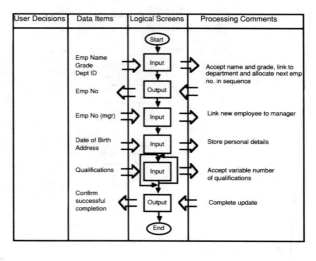

Fig. D.9

6.2

Figure D.10 shows the normalized set of relations for Scapegoat Systems and Fig. D.11 shows the optimized set.

Staff allocation sheet

UNF	1NF	2NF	3NF
Proj Code	Proj Code	Proj Code	Proj Code
Proj Desc	Proj Desc	Proj Desc	Proj Desc
Cust No.	Cust No.	Cust No.	Cust No.
Cust Name	Cust Name	Cust Name	
Staff No.			
Staff Name	Proj Code	Proj Code	Cust No.
Grade	Staff No.	Staff No.	Cust Name
No. of Days	Staff Name	No. of Days	
	Grade		Proj Code
	No. of Days	Staff No.	Staff No.
		Staff Name	No. of Days
		Grade	
			Staff No.
			Staff Name
			Grade

Fig. D.10

Invoice

UNF	1NF	2NF	3NF
Invoice No.	Invoice No.		
Date of Invoice	Date of Invoice	Invoice No.	Invoice No.
Cust No.	Cust No.	Date of Invoice	Date of Invoice
Cust Name	Cust Name	Total Cost	Total Cost
Cust Address	Cust Address	Cust No.	Cust No.
Proj Desc		Cust Name	
Start Date		Cust Address	Cust No.
Finish Date	Invoice No.		Cust Name
Man Days	Proj Code		Cust Address
Cost	Proj Desc	Invoice No.	
Total Cost	Start Date	Proj Code	
	Finish Date	Cost	Invoice No.
	Man Days	Man Days	Proj Code
	Cost		Cost
	Total Cost		Man Days
		Proj Code	
		Proj Desc	
		Start Date	Proj Code
		Finish Date	Proj Desc
			Start Date
			Finish Date

Fig. D.10 (continued)

Optimized relations

Invoice No.
Invoice Date
Customer No.
Total Cost

Customer No.
Customer Name
Customer Address

Invoice No.
Project Code
No. of Man Days
Cost

Project Code
Project Desc
Customer No.
Start Date
Finish Date

Project Code
Staff No.
No. of Days

Staff No.
Staff Name
Grade

Fig. D.11

6.3

Figures D.12 and D.13 show the relational data structure and the Composite Logical Data Design, respectively, for Scapegoat Systems.

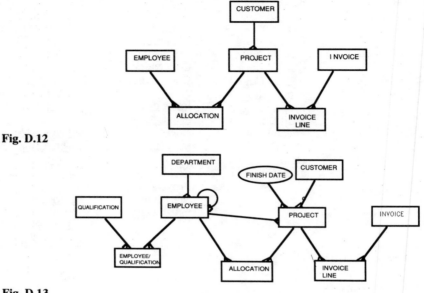

Fig. D.12

Fig. D.13

7.3

1.   Figure D.14 shows a portion of the Scapegoat Systems Composite Logical Data Design, showing the accesses and their sequence.

2.   Figure D.15 shows the Scapegoat Systems Logical Update Process Outline for the event Allocation of staff to project.

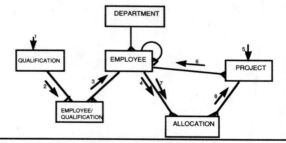

1,2,3	all employees with the required qualifications
4	check availability by looking at all current allocations
5,6	user enters project for allocation and system checks authority of project manager
7,8	Allocation occurrence created (these accesses could be the other way round)

Fig. D.14

Logical Update Process Outline

System: Scapegoat	Date:	Author:	Event: Allocation of staff to project

Batch / On-line	Access Type: (Acc Typ)	I - Insert D - Delete M - Modify L+ - add Link R - Read L- - remove Link	Read Path : (Rd Pth)	Dir - Direct Seq - Sequential Det - via Detail Mast - via Master

Op No.	Operation Description	I/O Ref	Acc No.	Entity	State Ind		Acc Typ	Rd Pth	Access Via	No. Acc	Data Items
					Val Pr	Set					
	Accept Qualification		1	Qual			R	Dir	Qual ID	1	Qual ID
	Display all staff with given Qualification		2 3	Qual/Emp Employee			R R	Mast Det	Qual Qual/Emp	50 50	Emp No., Name Grade
	Accept selection from user and display all Allocations for selected Employee. Repeat until available and suitable Employee found.		4	Allocation			R	Mast	Emp	5	Allocated Days/ Week
	Accept Project No. for required Project		5	Project			R	Dir	Proj No.	1	
	Authorise project manager by comparing with user ID		6	Employee			R	Det	Project	1	Emp No.
	Accept Allocation of Employee to Project		7	Allocation	–	1	I			1	Allocation details
	Link Allocation to Employee and Project		8 9	Employee Project			L+ L+			1 1	

Fig. D.15

Index